Taking Tinkercad to the Next Level

Enhance your ability to design, model, and 3D print
with one of the most intuitive CAD programs

Jason Erdreich

Taking Tinkercad to the Next Level

Copyright © 2024 Packt Publishing

All rights reserved. No part of this book may be reproduced, stored in a retrieval system, or transmitted in any form or by any means, without the prior written permission of the publisher, except in the case of brief quotations embedded in critical articles or reviews.

Every effort has been made in the preparation of this book to ensure the accuracy of the information presented. However, the information contained in this book is sold without warranty, either express or implied. Neither the author, nor Packt Publishing or its dealers and distributors, will be held liable for any damages caused or alleged to have been caused directly or indirectly by this book.

Packt Publishing has endeavored to provide trademark information about all of the companies and products mentioned in this book by the appropriate use of capitals. However, Packt Publishing cannot guarantee the accuracy of this information.

Group Product Manager: Rohit Rajkumar
Publishing Product Manager: Chayan Majumdar
Book Project Manager: Sonam Pandey
Senior Editor: Rashi Dubey
Technical Editor: K Bimala Singha
Copy Editor: Safis Editing
Indexer: Rekha Nair
Production Designer: Shankar Kalbhor
DevRel Marketing Coordinators: Nivedita Pandey and Anamika Singh

First published: September 2024

Production reference: 1200824

Published by
Packt Publishing Ltd.
Grosvenor House
11 St Paul's Square
Birmingham
B3 1RB, UK

ISBN 978-1-83546-800-5

www.packtpub.com

Contributors

About the author

Jason Erdreich is an educational leader, patented inventor, and all-around tinkerer and maker of things. With more than 10 years of experience in K-12 and higher education technology and engineering classrooms, Jason has worked to create engaging instructional spaces for students of all ages. Starting in 2017, Jason began to support learners outside of his own maker space via online content, training, product design, and collaborative work with companies such as LulzBot, Makera, and Autodesk. Through this experience, Jason has become a leader in educating learners of all levels in design thinking as they learn to utilize resources such as Tinkercad to create incredible things that can be manufactured successfully in almost any circumstance.

I want to thank the people who have always supported my projects and passions, including my students, my family, and my wife, Cara.

About the reviewer

Sharon Agun is a mathematics student at the University of Waterloo, specializing in statistics and actuarial science. With a strong analytical background, she has developed skills in data analysis and risk assessment, essential for understanding complex systems and making data-driven decisions. Sharon is also self-taught in various 3D software, exploring the intersection of technology and mathematics through projects in 3D design and printing. Passionate about innovation, she enjoys pushing the boundaries of what's possible with technology.

Table of Contents

Preface · xi

Part 1: Strategies for Successful 3D Modeling

1

Tinkercad, an Innovative Approach to 3D Design · 3

Technical requirements	4	3D Design	9
Creating in Tinkercad	5	Circuits	13
Who is Tinkercad for?	6	Codeblocks	14
Exploring 3D Design, Circuits, and Codeblocks	8	Summary	16

2

Tools and Strategies for Successful 3D Modeling · 17

Technical requirements	18	Using a mouse	25
Starting with a sketch	18	Keyboard shortcuts	26
Taking measurements	21	Summary	29
Choosing a device	23		

3

The Perspectives in 3D Design · 31

Technical requirements	32	2D orthographic views	38
Looking around	32	Summary	40
Orbiting through perspective views	35		

4

Designing through Constructive Solid Geometry   41

Technical requirements	41	Method 2	49
What is constructive solid geometry?	42	Manipulating shape parameters	51
Working in groups	45	Hiding and locking shapes	53
Method 1	48	Summary	56

Part 2: Advanced Tools and Features to Enhance our Designs

5

Creating and Manipulating Text Features   59

Technical requirements	60	Using individual characters	67
Typing and grouping text	60	Using text shape generators	68
Extrusions	62	Thinking outside the text box	71
Pockets	64	Summary	76
Orientation	64		

6

Using the Ruler and Workplane Tool to Dimension Our Designs   77

Technical requirements	78	Using the workplane tool	85
Working with the grid	78	Summary	89
Using the ruler tool	82		

7

Tools to Manipulate and Pattern Multi-Part Designs   91

Technical requirements	91	Duplicating and patterning shapes	100
Cruising	92	Summary	109
Aligning shapes	95		

8

Importing Models and Designs — 111

Technical requirements	112	Importing vector shapes	120
Tinkering from one design to another	113	Creating and importing custom vector image files	121
Importing 3D objects	116	Browsing the web for vector image files	124
Importing a 3D model	116		
Manipulating an imported 3D model	119	Summary	126

9

Making Our Own Shapes — 127

Technical requirements	127	Working with Shape Generators	137
Scribbling shapes	128	Creating custom shapes	140
Scribble shape basics	128	Summary	142
Putting the Scribble to use	131		

Part 3: Designing 3D Models for 3D Printing

10

An Introduction to 3D Printing and Production Techniques — 145

Technical requirements	145	Vat photopolymerization	154
What is 3D printing?	146	Stereolithography	156
How does 3D printing work?	149	Choosing the right material	157
Comparing 3D printing techniques	152	Printing with polylactic acid	158
Fused Filament Fabrication	152	Other commonly used materials	160
Fused Deposition Modeling	154	Summary	161

11

General Strategies for Creating Effective Models for 3D Printing — 163

Technical requirements	163	What is an overhang?	164
Avoiding overhangs	164	How can we avoid an overhang?	166

Creating segments, fillets, and chamfers	169	Designing for the first layer	175
		Utilizing surface area for better parts	176
Adjusting segments	170	Adding rafts and brims	178
Creating fillets and chamfers	171	**Optimizing the build plate**	**179**
Using CSG to create fillets and chamfers	174	**Summary**	**182**

12

Creating Tolerances for Multi-Part Designs 183

Technical requirements	183	Adding tolerances to our designs	188
What are tolerances?	184	Types of fit	188
How to calculate tolerances	186	Modeling different fits in Tinkercad	189
Determining tolerance based on printer type	186	Applying tolerances in a real-world setting	195
Determining tolerance based on material choice	187	**Summary**	**199**
Additional factors that may determine accuracy	187		

13

Design Mistakes to Avoid 201

Technical requirements	201	Working with resin-type 3D printers	210
Watching the workplane	202	Adjusting the performance of our walls	210
Identifying thin lines and walls	206	Connecting our parts	212
Working with extrusion-type printers	208	**Summary**	**214**

14

Exporting and Sharing Tinkercad Designs for Manufacturing 215

Technical requirements	215	3D printing Tinkercad designs with Autodesk Fusion	229
Exporting our designs	216	Sending Tinkercad designs to Fusion	230
Choosing and using CAM software	218	Preparing designs for 3D printing	232
Setting up Cura	219	Finding 3D printing services	241
Preparing a design in Cura	222	**Summary**	**243**

Part 4: Practical Applications, Start to Finish Designs to Test our Skills

15

Designing and Printing a Trophy — 247

Technical requirements	247	Creating a connection point	260
Designing the top part	248	Adding the recipient	262
Modeling the cup	248	Reorienting our parts	264
Modeling the post	250	Exporting and preparing for production	266
Finishing the cup part	254		
Designing the base part	257	Manufacturing the models	270
Creating a base platform	258	Summary	274

16

Fabricating a Multi-Part Storage Box with a Sliding Lid — 275

Technical requirements	275	Adding text features	286
Starting with the box	276	Adding image features	288
Cutting the grooves	279	Exporting and manufacturing our models	290
Making the lid	283		
Adding artwork	286	Summary	295

17

Modeling an Ergonomic Threaded Jar — 297

Technical requirements	297	Improving the lid part	312
Modeling the jar	298	Exporting and manufacturing the jar	314
Modeling the lid	302	Summary	318
Modeling the threads	304		
Adding ergonomic features	309		
Improving the jar part	309		

18
Building and Playing a 3D Puzzle — 319

Technical requirements	319	Adding artwork	331
Making the pieces	320	Preparing, exporting, and manufacturing the puzzle	336
Making the joint template	325		
Adding the joints	327	Summary	340

19
Designing and Assembling a Catapult — 341

Technical requirements	341	Creating the projectile system	353
Designing the frame	342	Completing the catapult frame	354
Starting with a Box shape	343	Making the projectile arm	356
Adding shapes to make our part	344	Making projectiles	358
Creating an assembly system	347	Manufacturing and prototype testing	359
Creating the assembly holes	347	Summary	364
Creating the Assembly Beams	350		

20
Prototyping a 3D-Printed Phone Case — 365

Technical requirements	366	Using the copy to create the opening for the phone	378
Acquiring the dimensions	366		
Modeling the phone	367	Adding aesthetic and ergonomic features	380
Rounding the corners	369		
Adding the ports and buttons	371	Adding a protective lip	380
Making the case	375	Adding unique design features	383
Making a copy of the phone model	375	Manufacturing with specialty materials	384
Creating the ports and buttons	376		
		Summary	389

Index — 391

Other Books You May Enjoy — 398

Preface

Tinkercad is one of the most exciting and intuitive CAD programs out there, and it is widely recognized for its user-friendly interface and versatility. While Tinkercad is commonly used by beginners to make basic 3D designs and things, Tinkercad also offers resources to create just about anything, from a coffee mug to a robot.

There are many other books available that are written for a beginner Tinkercad user, for example, someone who is looking to make simple things such as a keychain. This book is different as it not only offers an in-depth exploration of Tinkercad's 3D design features, but also creates connections to professional CAD and design techniques to equip you with the knowledge and skills for harnessing its full potential.

You'll start by enhancing your 3D design skills and diving into modeling topics and techniques in Tinkercad. You'll also learn fundamental tools for product design, such as technical drawings and measurement techniques, paving the way for modeling through efficient constructive solid geometry methods. Advanced Tinkercad modeling techniques, including the ruler and workplane tools, dimensions, patterns, shape generators, importing, and exporting are also covered. The book then focuses on translating your designs into real-world objects using 3D printing. You'll learn about common types of 3D printers, manufacturing tolerances, material selection, and practical applications with step-by-step guides for creating items such as threaded jars, puzzles, and phone cases.

I will be guiding you through this journey with a scaffolded approach toward learning 3D modeling in Tinkercad, and one with relevant connections to industry practices. My prior experience that allows me to do this comes from three main sources:

- More than 10 years of experience as a certified K-12 teacher and college professor for technology, engineering, and design
- Industry experience in developing content and products for 3D printing companies, software companies, and other rapid prototyping organizations
- My own interests, passions, and drive to create as an avid tinkerer and maker

It is my goal to pass my skills, knowledge, and passions onto you through this book so that you may experience the success and joy of bringing your own ideas to life, whatever they may be. By the end of the book, you will have the knowledge and skills needed to create intricate designs and models, all ready for successful production through 3D printing.

Who this book is for

If you are a student, hobbyist, tinkerer, or maker, who is familiar with the fundamental features of Tinkercad and looking to learn how Tinkercad can be used to create complex designs and models for 3D printing, then this book is for you! While this book looks at intermediate and advanced techniques for designing in Tinkercad, beginners striving to expand their abilities in CAD and learn more about 3D printing would also benefit. Even if you don't have a 3D printer of your own, that's OK too!

What this book covers

Chapter 1, Tinkercad, an Innovative Approach to 3D Design, looks at the general features and capabilities of the Tinkercad design application and what makes it so unique. We will also consider who Tinkercad is intended for and what Tinkercad can be used to create as we preview where this book will take us.

Chapter 2, Tools and Strategies for Successful 3D Modeling, provides an opportunity to develop skills and techniques to make modeling in CAD easier. We look at industry practices, such as technical drawings, and cover key tools and resources that are recommended for your workspace.

Chapter 3, The Perspectives in 3D Design, allows us to grapple with the challenges faced when designing in 3D. This chapter covers key concepts that we will continue to build on in the chapters to come, as well as tools in Tinkercad to support this challenge as we do.

Chapter 4, Designing through Constructive Solid Geometry, defines the concept of **constructive solid geometry** (**CSG**), the modeling technique on which Tinkercad is based. Throughout this chapter, we look at the tools and methods used to create, manipulate, and transform designs through fundamental CSG techniques.

Chapter 5, Creating and Manipulating Text Features, looks at the different ways to incorporate text into our Tinkercad designs. This includes the basic text feature, individual characters, text generators, and an example project to put your skills to the test.

Chapter 6, Using the Ruler and Workplane Tool to Dimension Our Designs, demonstrates the different measurement tools in Tinkercad. Through the tools introduced in this chapter, we will learn to incorporate dimensions into our designs to make our modeling more precise.

Chapter 7, Tools to Manipulate and Pattern Multi-Part Designs, dives deeper into the tools that can be used to manipulate shapes into complex designs. This includes learning how to combine shapes more effectively, and how to use tools in Tinkercad to automate our modeling through an example project.

Chapter 8, Importing Models and Designs, shows how models can be brought from one design to another, and the different features that allow us to import designs into Tinkercad, too. This includes importing 3D shapes and designs from other sources, as well as importing 2D images and artwork into the 3D space.

Chapter 9, *Making Our Own Shapes*, covers tools and shapes in Tinkercad that let us make our own unique shapes. After looking more closely at ways to draw and generate shapes, we look at how what we create can be turned into a custom shape for future use.

Chapter 10, *An Introduction to 3D Printing and Production Techniques*, offers an overview of what 3D printing is. We will not only look at how 3D printing works, but also identify common 3D printing techniques and uses as we begin to consider how we can manufacture our own models and designs.

Chapter 11, *General Strategies for Creating Effective Models for 3D Printing*, covers design features and strategies that can be used to make 3D printing our models more successful. We'll reference some of the tools and techniques discussed previously as we learn how to apply them specifically for 3D printing production.

Chapter 12, *Creating Tolerances for Multi-Part Designs*, covers the important topic of tolerances, something that must be considered in all forms of manufacturing. These concepts will elevate our skills in effectively designing multi-part models with the intention of manufacturing them using 3D printers.

Chapter 13, *Design Mistakes to Avoid*, looks at common mistakes that may cause our 3D models to fail during printing. Earlier, we looked at successful strategies to employ, but here we will instead analyze things to avoid doing and things to check before attempting to 3D print our projects.

Chapter 14, *Exporting and Sharing Tinkercad Designs for Manufacturing*, covers the steps needed to export designs from Tinkercad and prepare them to be manufactured using CAM. We look at different options for CAM software, as well as 3D printing services for those who may not have access to a 3D printer of their own.

Chapter 15, *Designing and Printing a Trophy*, challenges us to apply skills in creating multi-part models as we design a trophy in Tinkercad. We will discuss scale and the key concepts of CSG, as well as different techniques for 3D printing common projects.

Chapter 16, *Fabricating a Multi-Part Storage Box with a Sliding Lid*, challenges us to incorporate tolerances into our designs as we create a multi-part project. We utilize skills in adjusting our perspective and workspace as well as importing artwork to enhance our designs.

Chapter 17, *Modeling an Ergonomic Threaded Jar*, challenges us to not only consider aesthetics as we design our models, but ergonomics too. This project lets us apply previously learned skills and resources to automate and enhance our designs for the real world.

Chapter 18, *Building and Playing a 3D Puzzle*, combines many different topics as we are challenged to make the most complex project yet. This project challenges us to not only make an effective 3D model but also consider how an effective product can be designed and made, too.

Chapter 19, *Designing and Assembling a Catapult*, challenges us to utilize nearly all the tools and skills covered previously to create a fun toy! We consider efficiency in our design, as well as professional strategies for testing our prototypes as we complete this project.

Chapter 20, Prototyping a 3D-Printed Phone Case, challenges us to make one of the more complex projects as many factors must be considered to find success. We also consider different approaches to designing this real-world product, as well as different materials and techniques for manufacturing it.

To get the most out of this book

You will need a fundamental understanding of design software and computer principles, such as saving or opening files, to find success with this book.

Software/hardware covered in the book	Operating system requirements
Tinkercad	Windows, macOS, ChromeOS, or Linux
Cura	A web-enabled device with internet access
Fusion 360	A browser with the ability to run WebGL
Fused Filament Fabrication 3D Printers	
Fused Deposition Modeling 3D Printers	
Vat photopolymerization 3D printers	

In addition to covering Tinkercad, the focus of this book, we will also look at commonly used CAM programs to manufacture our Tinkercad designs with 3D printers. Cura, Fusion 360, and other CAM programs will be highlighted, but are not required to get the most out of this book. However, you will need access to a CAM program compatible with your 3D printer if you choose to 3D print your models as shown in this book. For users without a 3D printer, printing services available will also be covered in this book.

This book contains many long screenshots. These have been captured to provide an overview of various features. As a result, the text in these images may appear small at 100% zoom. However, you can take a look at the clear images at this link: `https://packt.link/gbp/9781835468005`.

Download the example models

Example models will be shared throughout this book to connect the text to the Tinkercad designs shown. After creating a Tinkercad account of your own, you can find these models on the author's Tinkercad page at `https://www.tinkercad.com/users/jvIiB20KFq0/`.

Conventions used

There are a number of text conventions used throughout this book.

`Code in text`: Indicates code words in text, database table names, folder names, filenames, file extensions, pathnames, dummy URLs, user input, and Twitter handles. Here is an example: "There are three sets of dots, one each for the `x`, `y`, and `z` axes of our design."

Bold: Indicates a new term, an important word, or words that you see onscreen. For instance, words in menus or dialog boxes appear in **bold**. Here is an example: "Instead, we see a slider to adjust the **Bevel** value, which allows us to put a beveled edge on the shape, as well as adjust the number of **Sides**."

> **Tips or important notes**
> Appear like this.

Get in touch

Feedback from our readers is always welcome.

General feedback: If you have questions about any aspect of this book, email us at customercare@packtpub.com and mention the book title in the subject of your message.

Errata: Although we have taken every care to ensure the accuracy of our content, mistakes do happen. If you have found a mistake in this book, we would be grateful if you would report this to us. Please visit www.packtpub.com/support/errata and fill in the form.

Piracy: If you come across any illegal copies of our works in any form on the internet, we would be grateful if you would provide us with the location address or website name. Please contact us at copyright@packt.com with a link to the material.

If you are interested in becoming an author: If there is a topic that you have expertise in and you are interested in either writing or contributing to a book, please visit authors.packtpub.com.

Share Your Thoughts

Once you've read *Taking Tinkercad to the Next Level*, we'd love to hear your thoughts! Scan the QR code below to go straight to the Amazon review page for this book and share your feedback.

https://packt.link/r/<1835468004>

Your review is important to us and the tech community and will help us make sure we're delivering excellent quality content.

Download a free PDF copy of this book

Thanks for purchasing this book!

Do you like to read on the go but are unable to carry your print books everywhere?

Is your eBook purchase not compatible with the device of your choice?

Don't worry, now with every Packt book you get a DRM-free PDF version of that book at no cost.

Read anywhere, any place, on any device. Search, copy, and paste code from your favorite technical books directly into your application.

The perks don't stop there, you can get exclusive access to discounts, newsletters, and great free content in your inbox daily

Follow these simple steps to get the benefits:

1. Scan the QR code or visit the link below

https://packt.link/free-ebook/9781835468005

2. Submit your proof of purchase
3. That's it! We'll send your free PDF and other benefits to your email directly

Part 1: Strategies for Successful 3D Modeling

In the first part of this book, we will be covering key concepts as we overview and identify strategies for successful 3D modeling. Starting with an overview of Tinkercad, we will quickly cover the key tools and resources needed to engage with the skills and activities that are to come in future parts of this book. This includes setting up your workspace, as well as building an aptitude in design thinking techniques and modeling in three dimensions. This part will also serve as an opportunity for all readers, regardless of prior experience, to work toward the level of proficiency needed to understand the advanced topics covered in future chapters for designing in Tinkercad and manufacturing our designs using 3D printing techniques.

This part includes the following chapters:

- *Chapter 1, Tinkercad, an Innovative Approach to 3D Design*
- *Chapter 2, Tools and Strategies for Successful 3D Modeling*
- *Chapter 3, The Perspectives in 3D Design*
- *Chapter 4, Designing through Constructive Solid Geometry*

1
Tinkercad, an Innovative Approach to 3D Design

Tinkercad launched in 2011 as an entirely new way to create 3D designs. Owned by Autodesk, one of the largest **computer-aided design** (**CAD**) companies in the business, Tinkercad now has over 80 million users worldwide. But unlike other CAD programs, Tinkercad allows users to create 3D models through an intuitive approach. Designing in Tinkercad feels more like working with toy blocks rather than traditional methods, but without losing the ability to create uniquely incredible designs.

Figure 1.1: A 3D design created in Tinkercad that resembles toy blocks

If you already have some experience in using Tinkercad to create 3D designs, this book is for you. In this chapter, we will identify key terms that you need to know in order to find success with Tinkercad. This includes identifying the key features of Tinkercad through the following topics:

- Creating in Tinkercad
- Who is Tinkercad for?
- Exploring 3D Design, Circuits, and Codeblocks

By the end of this chapter, you will not only know how to create unique 3D designs in Tinkercad but also begin to understand just what is possible through this innovative CAD program.

Technical requirements

To access and use Tinkercad, you must have a computer or tablet with internet access. Tinkercad is **web-based**, meaning that it works on nearly all devices through a web browser, but it also always requires an internet connection.

You will also need to create an account to access Tinkercad. Making an Autodesk account is free, or you can sign in using a supported account such as Google. Visit `www.tinkercad.com` to access Tinkercad.

After creating a Tinkercad account, the projects shown throughout this chapter can be found through the following links:

- `https://www.tinkercad.com/things/9UhmtgE9MxX-tinkercad-blocks`
- `https://www.tinkercad.com/things/71OAQjEcvOJ-elephant-zoo-model`
- `https://www.tinkercad.com/things/9y2ztmyuIit-roses-in-a-vase-example`
- `https://www.tinkercad.com/things/iuk0xmm6FP5-vise`
- `https://www.tinkercad.com/things/kPUp0PzIz3I-simple-sim-lab-catapult-example`
- `https://www.tinkercad.com/things/ltY5DPR74DR-toy-house-example-model`
- `https://www.tinkercad.com/things/8s9EUXuKcXU-sample-sensor-servo-circuit-with-microbit`
- `https://www.tinkercad.com/things/1LVeapRNLZu-electronic-prototype-project`
- `https://www.tinkercad.com/codeblocks/9hkP7aUUIQ2-vase-semana-13-al-17-sep`

Creating in Tinkercad

CAD has changed the way things are created across nearly every industry. Through a wide range of computer applications, professional designers and engineers use CAD software to create the products we buy, houses we live in, or even the movies we watch.

Tinkercad is different because it makes these opportunities available to everyone, including you! Through its free and web-based platform, users on almost any device in any location can use Tinkercad to create models, parts, simulations, circuits, programs, and more, all without needing a specific type of computer or expensive subscription.

But what also differs is what we can create. Many CAD programs serve one purpose, such as making an architectural model of a home or designing furniture. Tinkercad allows the user to make just about anything within a common space, as shown in *Figure 1.2* and *Figure 1.3*.

In *Figure 1.2*, we see that there are sidewalks, trees, animals, and concession stands presented in a 3D space:

Figure 1.2: A model of an elephant zoo

By combining basic geometric shapes, Tinkercad allows you to create complex designs by grouping and scaling different shapes together. We will dive deeper into these methods in *Chapter 4*.

As seen in *Figure 1.3*, we can also use Tinkercad to create everyday objects, such as a vase:

Figure 1.3: A model of a vase with flowers

The diversity of what's possible to be created in Tinkercad is one of its strongest features. And because of this, it is not possible to predict what you will be able to create using the knowledge and skills gained in this book, which I find to be incredibly exciting!

Tinkercad also allows you to make your 3D designs real by exporting models to be manufactured using production techniques, such as 3D printing. It's important to understand that there are limitations and constraints to this. Nearly anything is possible when working in a digital space, from flowers to elephants, but there are many mechanical constraints to consider once we look at manufacturing our designs. For example, you may want to make a large vase with a diameter of 200 millimeters, but your 3D printer is only 150 millimeters wide. Or you may find yourself looking to create small parts that are too delicate or detailed for your 3D printer to reproduce as they are designed in CAD.

Throughout this book, we not only consider how to create amazing designs in Tinkercad but also strategies for effective 3D modeling, which can be successfully manufactured using 3D printing production techniques, too. And because Tinkercad's unique design features and tools are suitable for a wide range of users, just about anyone can bring their ideas to life using Tinkercad.

Who is Tinkercad for?

The short answer is everyone. As mentioned, Tinkercad is an incredibly easy program to access making it suitable for a wide range of communities, schools, DIYers, and professionals alike. Because of the intuitive nature of the user interface, Tinkercad is often used by young creators who have never created with CAD before.

However, all users who are new to CAD could benefit from the simplicity of creation through Tinkercad's interface, regardless of age. Learning to design in 3D can be challenging at first, and using a simple program allows for all users to become proficient faster. Many also may think that Tinkercad can't be used for complex creations, which isn't the case.

As shown in *Figure 1.4*, you can use Tinkercad to create multi-part designs using specific **dimensions**, or measurements, using the **Ruler** tool:

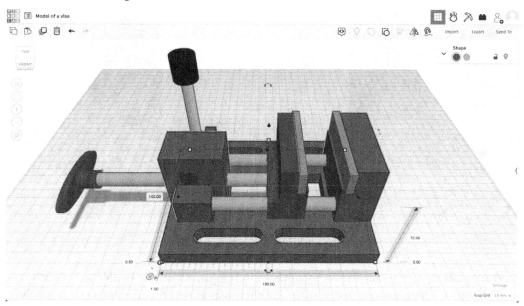

Figure 1.4: A multi-part vise with the Ruler tool shown

You can also import existing 3D designs or 2D images into Tinkercad, which allows you to use simple design features to make truly unique creations. This is our goal, and we will be covering these advanced techniques in detail throughout this book.

> **So who is Tinkercad really for then?**
> Not just beginners or young creators but everyone who is looking for a simple and intuitive way to bring their own unique ideas to life.

Tinkercad also allows you to share designs to edit or collaborate with another user in real time. As shown in *Figure 1.5*, users can press the **Invite** button to share a collaborative URL with a colleague:

Figure 1.5: Sharing a Tinkercad design

By sharing our designs, multiple users who may be working in a team will be able to view, edit, and apply feedback within a single design space. It's important to note that the link created is public, meaning that any user who obtains this link can not only view but also edit your design. You can also set your designs to be public, which will allow any Tinkercad user to search for and make copies of your designs.

So far, we've looked at some general features and capabilities for creating 3D designs and models in Tinkercad, but this is just one aspect of the program. We need to explore a little deeper to really consider what's possible using Tinkercad!

Exploring 3D Design, Circuits, and Codeblocks

It's important to identify that Tinkercad is really three different applications combined into one platform. You can access all three of these applications from the same website using the same account, and you will see the options to create within these three spaces by pressing the **Create** button (see *Figure 1.6*).

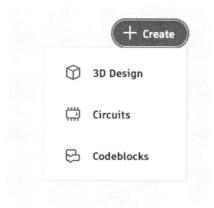

Figure 1.6: Options to create a new design in Tinkercad

In this section, you will learn about the differences between Tinkercad's three different applications, as well as when you might want to consider using one over the other. Two things that all three of these applications have in common are that they automatically save your work as you create, and your files are stored online in Tinkercad's cloud under your account.

3D Design

Everything we have discussed thus far has been focused on Tinkercad's **3D Design** application, and that remains the primary focus of this book. When using the **3D Design** application, we can combine different shapes to create just about anything in a collaborative 3D space.

In each following chapter, all references and topics will be focused on creating a 3D design in Tinkercad. There are also a number of different modes that can be used to interact with your 3D models and creations within the **3D Design** space. We will discuss them next.

Sim Lab mode

Sim Lab mode allows you to apply real-world physics and **material properties** to your 3D design. First, you must create a model in the **3D Design** application, such as the simple catapult shown in *Figure 1.7*:

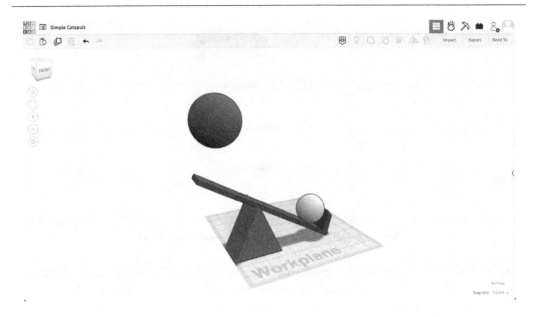

Figure 1.7: A multi-part catapult design with fulcrum, lever, weight, and projectile

Then, press the **Sim Lab** button to launch this mode after creating your design. You can select a shape you've created in the **Sim Lab** window to change what material it is made from. You can also set shapes to be **dynamic** (can move) or **static** (remain stationary). By pressing **Play**, gravity will be applied to your design. This allows you to see how multi-part designs may interact, balance, rotate, or move if they were in the real world, as shown in *Figure 1.8*:

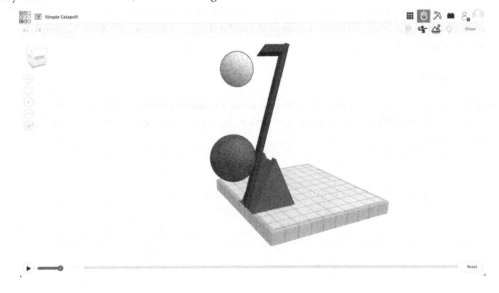

Figure 1.8: Simulating the motion of a simple catapult model using Sim Lab mode

In this example, the large sphere is set to be concrete while the smaller sphere is set to be polystyrene. Because of the difference in mass, the larger sphere causes the lever arm to rotate around the triangular fulcrum, which is set to **Static**, and as a result, the small sphere is thrown across our workspace when the parts collide. Sim Lab can be a powerful tool to test a design in a real-world space before manufacturing it, increasing your ability to design complex models more successfully.

Another mode to use in the **3D Design** space is called **Bricks mode**.

Bricks mode

Like Sim Lab, we first must create a model in the **3D Design** application before pressing the **Bricks** button to enable this mode. By pressing the **Bricks** button, your 3D design will be recreated using plastic bricks, as shown in *Figure 1.9*:

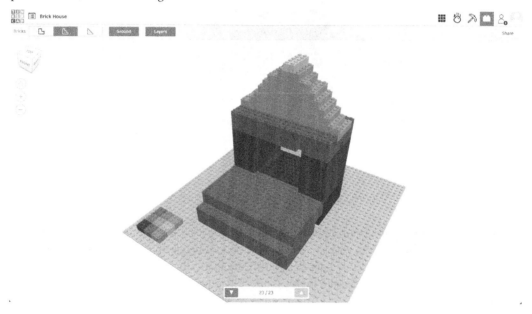

Figure 1.9: Viewing a 3D model in Bricks mode

You can set the **resolution** of the model, which will change how many bricks are used to create your shapes. Higher resolutions will use more bricks but also create a more detailed design. You can also view your design in **layers** to show you how your design can be built using these plastic bricks in the real world, one step at a time.

Similar to **Bricks** mode, we have **Blocks mode**.

Blocks mode

This mode allows you to **render** a 3D design in a different workspace. As with **Sim Lab** mode and **Bricks** mode, you first need to create a model in the **3D Design** application before viewing it in **Blocks** mode, as shown in *Figure 1.10*:

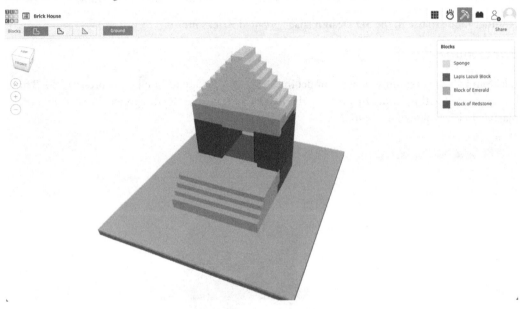

Figure 1.10: Viewing a 3D model in Blocks mode

By pressing the **Blocks** button, your 3D design will be redrawn as if were a model from the *Minecraft* video game. Like the **Bricks** view, you can set the resolution to change the detail of your design render within this mode. The colors used in your design will automatically be converted to blocks available from Minecraft, which can be adjusted in this mode. This mode may be a powerful feature to take advantage of for creators looking to use Tinkercad to create models for video games.

In addition to being able to make 3D models that might be suitable for unique projects, Tinkercad's additional design spaces may allow you to do the same to create electronics or computer programs!

Circuits

The **Circuits** application differs from the **3D Design** application as it is intended to create and simulate **electronic circuits**. As in **3D Design**, you can use a drag-and-drop interface to interact with **Circuits**. However, the tools available and intention for this application are vastly different, as shown here:

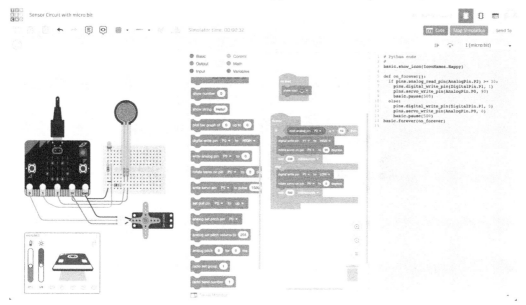

Figure 1.11: A micro:bit circuit simulation created using Tinkercad Circuits

As seen in *Figure 1.11*, electronic components such as wires, resistors, sensors, motors, and LEDs can be combined with **microcontrollers**, such as the *micro:bit*, in a digital space. **Circuits** allows you to create both analog and digital electronic circuits, as well as write code in either a block-based or script-based language before simulating how your circuit may function if it were built in the real world.

The **Circuits** application may allow you to design and **prototype** the entirety of your project within Tinkercad. Consider if you were designing an electronic prototype that used a micro:bit, servo motor, and LED, as shown in *Figure 1.11*. You could first design and test the electronic element of this project in the **Circuits** application before moving into the **3D Design** application, as shown here:

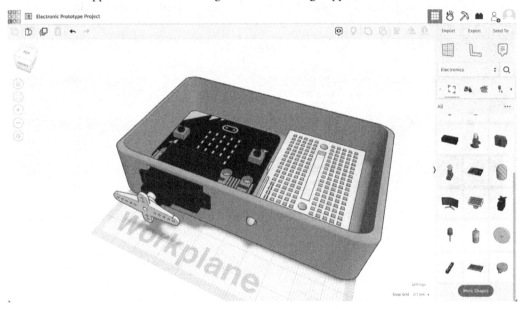

Figure 1.12: Creating a 3D design with electronic components

In the **3D Design** application, many of the electronic components available in **Circuits** are included as shapes from within the **Electronics** category, as shown in *Figure 1.12*. These shapes can be selected and moved around just like normal shapes, but their size is locked to keep them true to **scale** for the real components they represent. Using these shapes could allow you to make a 3D printed enclosure for your circuit design manufactured using a 3D printer, all by using Tinkercad's different applications.

Codeblocks

Codeblocks again differs from **3D Design**, though the outcome is very similar. In the **Codeblocks** application, you can create and manipulate 3D shapes to make models as you might in the **3D Design** space, but this is done entirely using a **block-based programming language**.

Every single step taken to create a model would be done through a unique block of code. You can combine blocks of code to create a **program**, and even use looping or repeat functions to make a complex geometrical design, as shown here:

Exploring 3D Design, Circuits, and Codeblocks 15

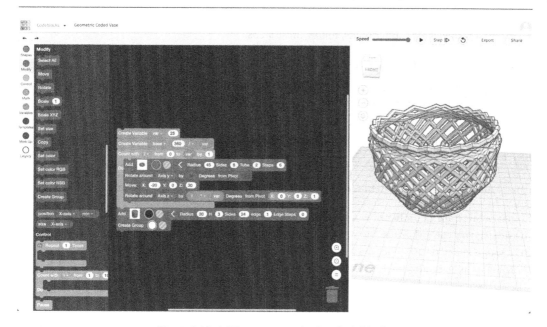

Figure 1.13: A 3D vase created using Codeblocks

The vase created in *Figure 1.13* was made using the code blocks, also shown in this figure. Individual commands for creating shapes and setting size, position, color, and rotation were all combined to create this design. If you are looking to not only learn how to create 3D designs but also to gain abilities in computer programming, **Codeblocks** may be a suitable Tinkercad application for you.

We can also export the models created in **Codeblocks** into the **3D Design** application, or even download them directly for manufacturing using 3D printing, as shown in *Figure 1.14*:

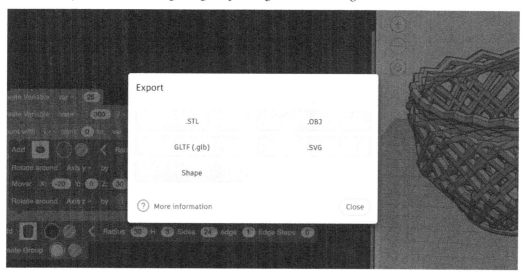

Figure 1.14: Options for exporting models made through Codeblocks

To export a model from **Codeblocks**, follow these steps:

1. Press the **Export** button after creating a model in **Codeblocks**.
2. If you want to download for 3D printing, select one of the file formats shown in *Figure 1.14*. To bring this model into the **3D Design** application, select **Shape**.
3. Complete the details in the window that appears by naming the shape, providing a description, and setting the parameters for your design. Then, click **Save Shape**.
4. From the main Tinkercad window, press **Create** and create a new **3D Design**.
5. You can find your **Codeblocks** model under the **Shape** category titled **Your Creations**.

To recap, there are a number of ways to design and create in Tinkercad. If you are looking to create a 3D model, you can either use the principles of constructive solid geometry to create a model in the **3D Design** space or use block-based coding in **Codeblocks** instead. Within the **3D Design** space, there are tools to render your models in different modes, such as **Sim Lab** or **Bricks**.

Tinkercad also provides tools for designing electronic circuits and computer programs using the **Circuits** space. While this may not be the primary focus of this book, it is a powerful feature to encompass the entirety of an electronic prototype solution you may be creating from within a single CAD application.

Summary

In this chapter, you have begun to see what can be created using Tinkercad. Through Tinkercad's intuitive and easily accessible interface, users of all levels can create a wide range of 3D designs.

Tinkercad also has two other applications, **Circuits** and **Codeblocks**. While these applications differ from the **3D Design** application covered throughout this book, they too can be used to create unique things, including 3D models and prototypes.

While Tinkercad is often used by beginners to create simple 3D models, we will be focusing on how advanced users can also take advantage of Tinkercad's intuitive interface to make complex multi-part designs both in a digital space and in the real world using technology such as 3D printing.

When you're ready, turn to the next chapter to learn about important tools and strategies needed to begin to make advanced 3D models using Tinkercad.

2
Tools and Strategies for Successful 3D Modeling

As we work to create complex designs with multiple parts that may be produced using 3D printing technology, we must also consider steps and strategies that we can employ to be more efficient in our creation, and successful with production. As you progress through the first part of this book, you will be introduced not only to the possibilities of creation using Tinkercad but also to strategies that may bring you greater success. Learning the fundamentals of designing in CAD is a key step for any successful designer, modeler, or creator.

In this chapter, you will be introduced to design strategies, techniques, and tools that may support you as you create complex 3D models using Tinkercad through the following topics:

- Starting with a sketch
- Taking measurements
- Choosing a device
- Using a mouse
- Keyboard shortcuts

By the end of this chapter, you will have identified a workflow that suits your design style and needs, as well as steps that you can take to create a workspace that is more conducive to advanced creation through computer-aided design. This will allow us to engage with more advanced techniques, such as creating complex shapes and multi-part models suitable for 3D printing later in this book.

Technical requirements

In this chapter, we will look at tools and resources that are not required but recommended to find greater success through advanced CAD techniques. We will look at a range of devices that are well suited for Tinkercad, all of which must have the ability to run **WebGL** in the browser. Visit `https://get.webgl.org/` to determine whether your device is compatible.

Having a computer mouse rather than a touchpad may increase the ease of use when working with Tinkercad. A touchscreen-enabled device may also allow for a more intuitive workflow for younger users, or for users with fine motor disabilities.

You also may need a pencil or writing utensil, as well as paper or a sketchbook to collect your thoughts and brainstorm your designs. Rulers or calipers are also key tools to take detailed measurements of parts or components for your designs.

Additionally, the example models shown in this chapter can be obtained through the following links:

- `https://www.tinkercad.com/things/2pfTo63nud2-ar-chair-model`
- `https://www.tinkercad.com/things/hUhDqZcK2L7-example-gear-model`

Starting with a sketch

Many users find designing in 3D to be challenging, especially when looking at 3D designs on a 2D computer screen. As we think of ideas and features for our 3D designs, it is sometimes faster and more effective to brainstorm off the screen.

Sure, sketching can be a product or an art form in itself, but it's also one of the earliest forms of communication and one of the core principles of early CAD programs. Even in today's modern age, real-world designers, engineers, and modelers use sketching as a tool for brainstorming, planning, and communicating their ideas before diving into CAD software.

You don't need to be an excellent artist to create an effective sketch. In fact, **rough sketching** is an incredibly effective way to come up with many ideas quickly before diving into a 3D model. Let's say that I am designing a phone case, a product which has been created many times. I might make **thumbnail sketches** to brainstorm many ideas through small rough sketches before combining my favorite ideas into one final sketch, as shown in *Figure 2.1*:

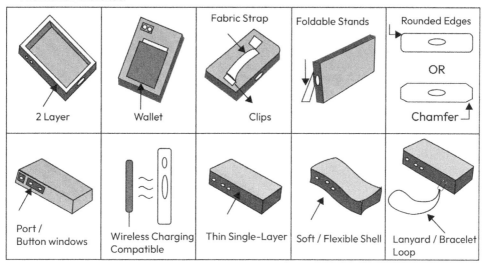
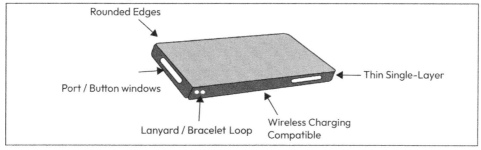

Figure 2.1: Using sketching techniques to brainstorm a phone case design

The small thumbnail sketches shown in *Figure 2.1* are a long way away from what I would consider to be art, but they allow me to get my ideas visualized far more quickly than I could have done in even the simplest CAD program such as Tinkercad.

After being able to see your ideas in front of you, you could share them with a colleague for feedback or take notes on your favorite features quickly. After these brainstorming steps, you can then narrow your ideas down into a design that is worth putting in the time to model using CAD. This is an effective technique that is really used by professionals, though there are ways to refine and improve our sketches to make them a bit more effective.

Let's look at the **technical drawing** shown in *Figure 2.2*:

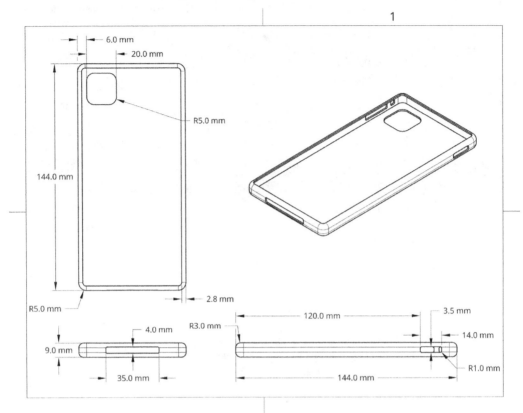

Figure 2.2: An example of a technical drawing for a phone case

A technical drawing is a sketch that often has multiple views, in addition to key information as to how a part or product should be made. When looking at the drawing in *Figure 2.2*, we can see a 3D drawing in the top right corner. We call this an **isometric** drawing, as it looks at the part from a 30-degree perspective view, which makes it appear to be 3D. This provides us with a real-world visual of how this part may appear if it were sitting in front of us.

We can also see (in *Figure 2.2*) three 2D sketches showing the same part from different views. This is called an **orthographic** drawing. For orthographic drawings, we typically sketch the top, front, and right sides of a part, though that can vary based on the detail or complexity that you are trying to convey.

While the orthographic drawing doesn't quite show us what the part looks like in the real world like the isometric view does, it does provide clear details about the different sides of our design, which can often be lost in an isometric drawing.

One of the most important features of a technical drawing is **dimensions** or measurements. Typically, we only place dimensions on the orthographic drawing. We can also share dimensions across these views, as they are all the same part. By combining dimensions, isometric, and orthographic drawings into a single document like in *Figure 2.2*, we can communicate a lot of information and detail about our parts clearly to ourselves, peers, or even a production team. However, if we want to annotate our sketches with dimensions, we must also gain proficiency in taking measurements accurately.

Taking measurements

As shown in *Figure 2.2*, dimensions are an important component of the design process when we are working to design products that need to fit together in the real world. Before you dive into creating a design, it is best to obtain all the dimensions to incorporate into your models. **Rulers** and **calipers** are key tools to have on hand to obtain these measurements.

Imagine that you are making a box or case that must fit something, such as a phone. We would need to know its width, length, and height, as well as the placement of features such as buttons or ports. Using a tool such as digital calipers, as shown in *Figure 2.3*, makes finding these measurements a bit easier and more accurate.

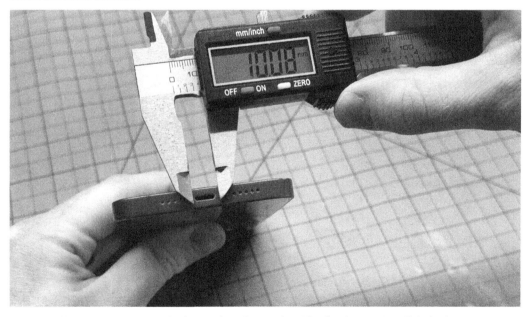

Figure 2.3: Measuring the button location on the side of a phone using digital calipers

You can also utilize your sketches or drawings to organize all these measurements effectively, as shown in *Figure 2.2* earlier.

However, you may often find that key dimensions for common products such as phones or connectors are readily available from the supplier. Take a moment to search on the internet for a specific model phone, plus the word *dimensions*, and see what sorts of images come up. You may find a detailed technical drawing for that device.

Industrial parts suppliers and retailers will often provide drawings for the parts that you may want to include in your designs, such as screws, bolts, springs, or hinges. The technical drawing shown in *Figure 2.4* provides orthographic and isometric views of a bolt, as well as dimensions that can be used in a design.

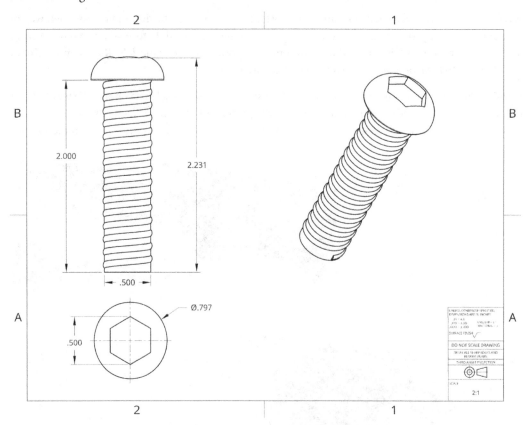

Figure 2.4: A technical drawing provided for a bolt

Gaining skills in sketching will not only allow you to create a sketch to speed up your design process but also enable you to read and utilize existing technical drawings such as the one shown in *Figure 2.4*. These skills are not only useful as you start to brainstorm and create a new project but can also come in handy later in the design process as you are working in Tinkercad to create a 3D model.

In addition to being able to source drawing files and dimensions for components and parts, you may also be able to find 3D models that can be imported directly into your Tinkercad design, as shown in *Figure 2.5*:

Figure 2.5: Importing a 3D file of a bolt into a Tinkercad design

In *Chapter 7*, we will learn how to use the ruler tool to apply accurate dimensions to our 3D models based on the sketches and technical drawings that we are learning to use in this chapter. Later, in *Chapter 9*, we will learn how to import 2D and 3D files to enhance our own designs, as shown in *Figure 2.5*.

As you prepare to develop your models using the sketches that you create and the dimensions that you record, it's also important to choose a device that can support your personal preferences and needs. Fortunately, Tinkercad is compatible with a wide range of devices.

Choosing a device

We've already learned that Tinkercad works on nearly any device if you have access to the internet, though you may find that some devices offer different features that you can take advantage of.

When working with a touchscreen-enabled device, Tinkercad supports a drag-and-drop interface right in the 3D design editor. You can select, drag, and manipulate shapes using your finger or a stylus just like you would with a mouse pointer. You can even rotate your view around by dragging in the 3D space, making the navigation feel more intuitive. This direct interaction approach created by a touchscreen-enabled device tends to support the needs of younger creators or creators with fine motor disabilities. If you're using the Tinkercad app on an iPad, there's another interaction feature available in addition to a touchscreen-enabled interface: **AR Viewer**.

AR Viewer in the Tinkercad app for iPad allows you to place your 3D models in a real space using AR. As shown in *Figure 2.6*, a 3D model of a chair created in Tinkercad appears to be placed in the living room of a home using AR viewer in the Tinkercad app for iPad:

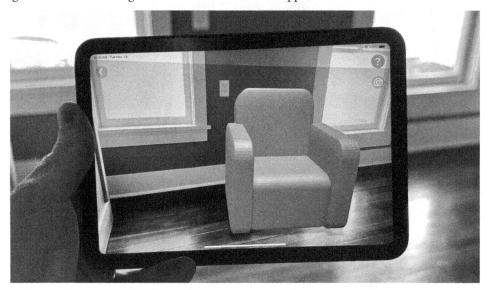

Figure 2.6: Viewing a chair model in a real-world space using AR viewer for iPad

While touch-enabled and augmented reality features offer an enhanced Tinkercad experience, they are not necessary to use Tinkercad to create incredible things. Tinkercad is known to work well on Windows, Apple, and Chromebook devices in the Chrome, Safari, and Edge web browsers. The only strict requirement your device must have to run Tinkercad is the ability to support WebGL in the browser.

Working with touchscreen devices such as Chromebooks or tablets may offer an intuitive design approach that works for your personal preferences. However, I personally find that there is another approach to designing which offers more control.

Using a mouse

While it is not required, I recommend that you consider using a mouse with whatever device you choose to work on. You may find that using a mouse over a touchscreen or trackpad is not only easier and more precise, but also allows for more effective navigation and control.

As shown in *Figure 2.7*, a standard two-button mouse offers some additional levels of control for working in Tinkercad's 3D space:

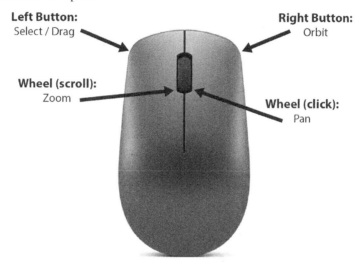

Figure 2.7: The Tinkercad control buttons when working with a mouse

The left button allows you to click, drag, and select things, while the right button allows you to **orbit** around in the 3D space. Orbiting is the action of rotating around your design to view it from every angle, which is something that can also be achieved by dragging the *view cube* in the top-left corner of your window.

Scrolling the wheel of a mouse forward and backward will zoom your view in and out. If you are in **perspective view**, you will zoom to where the mouse pointer is located on the screen. If you are in **Flat View (Orthographic)**, you will always zoom toward the center of your screen. We can change between these views using the sidebar button, as shown in *Figure 2.8*. By clicking and holding the wheel of the mouse like a button, you can pan your view from side to side.

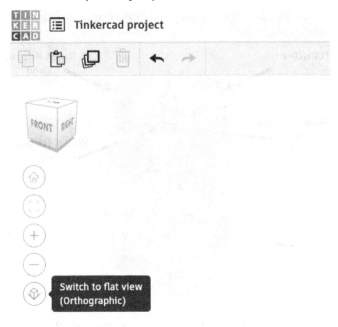

Figure 2.8: Changing the view perspective in Tinkercad

The most important thing is to create a workspace that is comfortable for you. No one strategy or device is necessarily better than the other, as long as you feel comfortable with the tools you have available so that you may use them to create effectively. In addition to considering using touchscreens or mice in your workspace, you may also find that devices with a keyboard can increase your comfort and streamline your workflows.

Keyboard shortcuts

Another strategy for success is to take advantage of Tinkercad's **keyboard shortcuts**, as outlined in *Figure 2.9* (image credit – `http://www.tinkercad.com`):

AUTODESK Tinkercad keyboard shortcuts — PC / Mac

Ctrl = Command   Alt = Option

Shape properties

Group	Ctrl + G
Ungroup	Ctrl + Shift + G
Make a Hole	H
Make a Solid Color	S
Make Transparent	T
Lock or Unlock	Ctrl + L
Hide	Ctrl + H
Show all	Ctrl + Shift + H
Group	Ctrl + G

Helpers

Place Ruler	R
Place Workplane	W
Place Workplane at shape	Shift + W
Show Shape Workplane	E

Viewing 3D space

Fit selected in view	F
Orbit	Right mouse or Ctrl + Left mouse
Pan	Middle mouse or Shift + Right mouse
Pan	Ctrl + Shift + Left mouse
Zoom in or out	+ and − or Scroll

Commands

Copy	Ctrl + C
Paste	Ctrl + V
Duplicate, repeat duplicate	Ctrl + D
Drag a copy	Alt + move shape
Undo	Ctrl + Z
Redo	Ctrl + Y
Redo	Ctrl + Shift + Z
Select all	Ctrl + A
Select multiple	Shift
Drop to workplane	D
Align	L
Mirror or flip	M

Move, rotate, and scale shapes

Move 1 grid step (X/Y axis)	Arrow keys
Move up (Z axis)	+ Ctrl
10x move	+ Shift
Rotate snap to 45°	Shift + Rotate handle
Scale about center	Alt + Scale handle
Uniform scale	Shift + Scale handle
Cruise tool	C
Cruise below surface	Shift while cruising

Printable PDF version: autode.sk/tinkercad-keyboard-shortcuts

Figure 2.9: Keyboard shortcuts for working in Tinkercad's 3D design application

In general, there is an on-screen button for almost all of the keyboard shortcuts shown in *Figure 2.9*. This means that you do not need to know these shortcuts to find success when creating in Tinkercad. However, learning and using the keyboard shortcuts may increase the speed at which you are able to create, as well as your effectiveness.

Let's say that you're creating a complex design with multiple shapes and parts. You would need to move your mouse to three different locations on the screen to press the on-screen buttons each time you want to add, duplicate, and group shapes. If you make a mistake, it will take two more mouse movements to undo something. Alternatively, if you were to take advantage of keyboard shortcuts, your mouse could stay in the design part of the window while your fingers perform all the manipulative actions via the keyboard.

There's also a handful of shortcuts that allow you to do things for which there isn't a button. For example, pressing *D* on your keyboard drops selected shapes that are floating so that they touch the workplane, as shown in *Figure 2.10*:

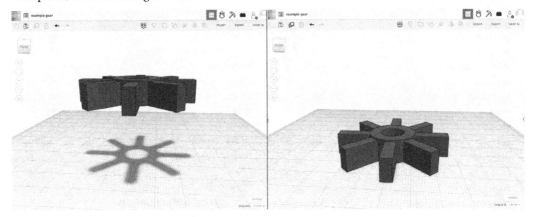

Figure 2.10: An object floating above the workplane (left) is precisely lowered to the workplane (right) using the drop shortcut

Using the arrow keys allows you to move objects more precisely than dragging them with your mouse does. Each time you tap an arrow key, your shape will move in the *x* or *y* direction by an amount that is based on the distance set using the **Snap Grid** menu in your design space, as shown in *Figure 2.11*:

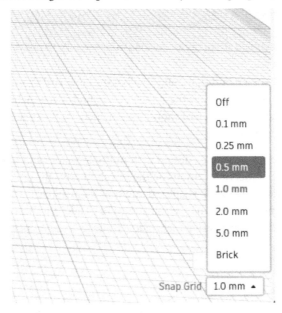

Figure 2.11: Adjusting the snap grid increments in the 3D design application

If you hold the *Ctrl* key while pressing the *up* or *down* arrows, your shape will move in the *z* direction. You can adjust the snap grid increments to make arrow key movements smaller or bigger based on your design needs or personal preferences.

Additionally, some keyboard shortcuts can be combined with mouse movements to enhance how we manipulate shapes, such as holding the *Shift* key. You can hold the *Shift* key while selecting shapes to be able to select more than one shape at a time. You can also hold *Shift* when dragging the corners of a shape to change its size; this will scale the shape uniformly in all dimensions. If you hold *Shift* while rotating an object, as shown in *Figure 2.12*, the object rotation will be locked to 45-degree increments.

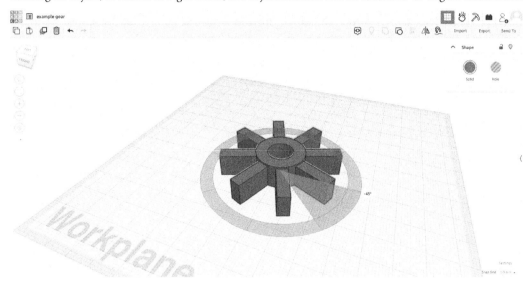

Figure 2.12: Rotating a shape while holding the Shift key

Tinkercad's keyboard shortcuts not only allow you to complete actions more quickly, such as copying a part or reversing an action, but they also enhance your capabilities and control. Learning the shortcuts to transform objects, such as rotating an object or moving it across the screen, will allow you to increase your precision and efficiency in designing complex models.

Summary

In this chapter, you learned that there are many different approaches to developing 3D models and designs using CAD, as well as many different tools that can be used to find greater success. While no one technique may be better than the others, you may find that a combination of tools and strategies feels more comfortable for you, as you learned in this chapter.

Throughout this chapter, you learned how you can design more efficiently by creating sketches to brainstorm and communicate your ideas. You also learned about the importance of dimensions and how to take accurate measurements of parts and components to include in your models. As you transition into designing in Tinkercad, you should consider which device or tools make you feel comfortable as you gain proficiency in computer-aided design.

As you progress through this book and learn how to create complex designs, consider the strategies and techniques introduced in this chapter to assist you in finding greater success and increasing your efficiency in using Tinkercad to create unique things.

When you're ready to learn how to apply some of these strategies in Tinkercad's 3D space, turn to the next chapter to dive into the different perspectives within 3D design.

3
The Perspectives in 3D Design

One of the most challenging aspects of learning CAD is the sensation of creating a 3D design on a 2D screen. Tinkercad is no different, and in this chapter, we'll look at the key skills and strategies that will allow you to gain proficiency, overcome this challenge, and create complex 3D models more effectively. This chapter will also allow us to put some of the shortcuts and advanced navigation techniques that were introduced in *Chapter 2* to work as we engage with new concepts.

In this chapter, we will cover the following topics:

- Looking around
- Orbiting through perspective views
- 2D orthographic views

Throughout this chapter, you may notice some terms that were previously introduced in *Chapter 2* when we learned how to brainstorm our designs through technical drawing and sketching techniques. After all, CAD was created after general principles for technical drawing were well established, meaning that similar terms and methods used in professional sketching are often found in CAD programs as well. While Tinkercad presents these concepts in intuitive methods that are often simpler than alternative CAD programs, professional strategies are still applicable as we strive to create complex 3D designs using Tinkercad.

By the end of this chapter, you will not only be comfortable with tools, shortcuts, and buttons within the Tinkercad interface that you may not have known about before, but also more proficient in designing models using Tinkercad's 3D design application.

Technical requirements

While it isn't required, using a keyboard and mouse rather than a touchscreen or trackpad is strongly recommended. Throughout this chapter, shortcuts and tools will be referenced using this recommended setup. Additional device and workspace recommendations can be found in the topics covered in *Chapter 2*.

The example models shown in this chapter can be obtained through the following links:

- `https://www.tinkercad.com/things/abzTOmBPrWo-perspective-mug-example-from-chapter-2`
- `https://www.tinkercad.com/things/0SOPSZAZinB-keychain-example`
- `https://www.tinkercad.com/things/71OAQjEcvOJ-elephant-zoo-model`
- `https://www.tinkercad.com/things/1LVeapRNLZu-electronic-prototype-project`

Looking around

One of the most important concepts we need to cover is that the entirety of your design cannot be viewed from a single perspective. For example, let's look at a coffee mug model designed in Tinkercad:

Figure 3.1: Looking at a 3D coffee mug from a front perspective view

From the front view shown in *Figure 3.1*, everything looks good. We can see that there is an opening where the coffee should go and a handle attached to the side. But we are looking at this through a 2D screen (or book), and this model is actually 3D. As a result, we need to constantly change our views to thoroughly inspect our designs; otherwise, we might miss something, as shown in *Figure 3.2*. This is the same coffee mug model, but we are now looking at it from the front-right side:

Figure 3.2: Looking at a 3D coffee mug from a front-right perspective view

From this angle, we can see that the handle is not attached, and it's not even really a handle at all but instead a complete Torus shape. If we designed this entirely from the front perspective, which is something that might feel natural at first, we would have missed this mistake entirely. In addition to **orbiting** around objects to see them from multiple sides, we should also look up and down; otherwise, we may miss something else, as shown in *Figure 3.3*:

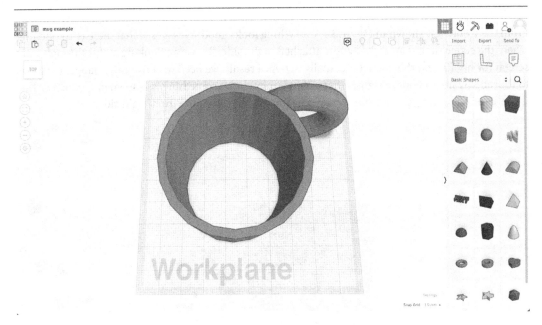

Figure 3.3: Looking at a 3D coffee mug from a top perspective view

Again, we are looking at the same coffee mug design shown in *Figure 3.1* and *Figure 3.2*, but we are now looking down on it from the top. It's now clear that this really isn't a mug at all, and instead just a tube with no handle or bottom to hold your coffee.

Learning from this example, we need to change our view to inspect our models throughout the design process. After adding a shape, look around. After grouping shapes, look around. Constantly change your view as you're designing to inspect and ensure that your model is being created how you intend it to be. Another common mistake that can be missed if we don't look around is creating models with objects that are floating, as shown in *Figure 3.4*:

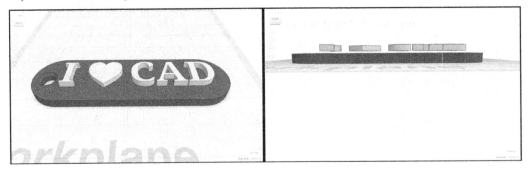

Figure 3.4: A model of a keychain shown from two different perspective views

In the view shown to the left of *Figure 3.4*, we are looking at a top-front perspective. Here, everything looks fine, as if the text is touching the box below it to create a solid model. But in the view shown to the right of *Figure 3.4*, we can see that the text is floating above our lower base shape.

I'm pointing this out because Tinkercad will still allow you to group shapes, even if they aren't touching. As a result, it is very easy to think that you have created a solid part while gaps and spaces exist, often ones that are far smaller than the examples shown in this chapter. If we were to export and 3D print the mug or keychain model shown, we would run into catastrophic issues at the production stages of these projects, as discussed later in *Chapter 13*.

As a result, it is important to consistently look around your models and designs, as well as gain proficiency in how to do this. We'll discuss this in the next section.

Orbiting through perspective views

We can change our view in a few different ways. **Orbiting** is the process of moving around the model in a 3D path. You can orbit by clicking and dragging the **view cube** in the top-left corner of the design window, as shown in *Figure 3.5*:

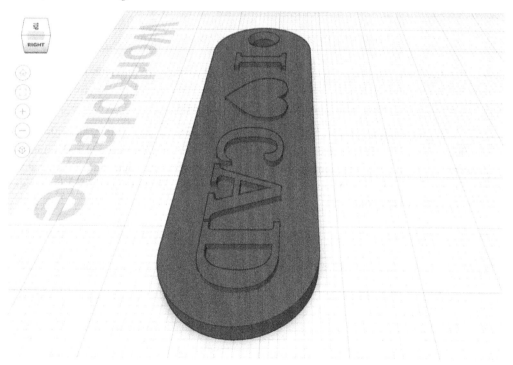

Figure 3.5: Using the view cube to change the view of a 3D model

The view cube always refers to the angle at which you're currently viewing your model. You can also click on any side or edge of the view cube to snap to a specific view. For example, if we wanted to look at the top right-hand side of our model, we could click on the top-right edge of the view cube, as shown in *Figure 3.5*.

If you're using a mouse, you can right-click and drag the cursor anywhere in the design window to orbit in the same way we can click and drag the view cube. We can also **pan** using a mouse by clicking the scroll wheel and dragging across our design window. Panning doesn't move in a 3D path like orbiting, but instead will change our view from side to side in whatever angle we are looking at. This is especially useful when we're zoomed in on a design and looking at it from close up, as shown in *Figure 3.6*:

Figure 3.6: Panning across a model from left to right

Between the three views shown in *Figure 3.6*, we can see that we move to the right in each view. This was done by clicking and holding the scroll wheel of a mouse while dragging our mouse pointer across the design window to the left, which panned our view to the right.

As you're designing complex shapes in Tinkercad, it's important to get comfortable with orbiting and panning around your designs to ensure you don't miss any details. **Zooming** in and out is also a key strategy to assist in designing. Sometimes, it is difficult to be precise when you're looking at your model from a far-out view. You can zoom in by pressing the + and – buttons on the left-hand side of your screen, as shown in *Figure 3.7*:

Orbiting through perspective views 37

Figure 3.7: Zooming in and out using the on-screen buttons

Whenever you press the zoom buttons shown in *Figure 3.7*, you will zoom toward the center of your window. You can also press the + and – buttons on your keyboard or use the scroll wheel on a mouse to zoom in and out as well. If you use the scroll wheel on a mouse to zoom, you will zoom to where your mouse pointer is located unless you're in an orthographic view, as discussed in the following section.

Pressing the **Home** button above the zoom controls shown in *Figure 3.7* will reset your view to the initial front-top 3D perspective view we see when Tinkercad loads a design. If you press the **Fit all in view** button, or *F* on your keyboard, your view will automatically zoom out so that the entirety of your design fits within your design window.

2D orthographic views

By default, the Tinkercad 3D design application shows our models in a 3D perspective view. This is useful as it makes our designs appear to be 3D, even though we're viewing them on a 2D screen (or book). But sometimes, this isn't useful and instead makes designing rather difficult. In *Figure 3.8*, we're looking down on a model from the top perspective view and some of the details are hard to see:

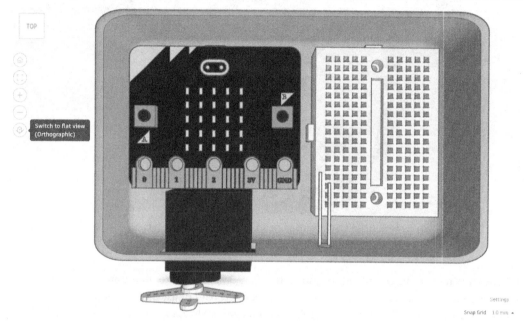

Figure 3.8: Looking at the top of a model in a 3D perspective view

Because we're looking through the default 3D perspective view, the shapes appear to be 3D. But we're looking at the top of an object, which should be a 2D perspective view. Because of this, the shapes tend to look misaligned or warped, sometimes making it difficult to see all the details in our design or model effectively. This is like the **orthographic** and **isometric** concepts we discussed in *Chapter 2*.

When we're trying to design our models from a 2D view, which can be handy when we're creating complex designs or placing specific measurements between shapes, it's important to disable the perspective view and enable the orthographic view for a more accurate representation of our design, as shown in *Figure 3.9*:

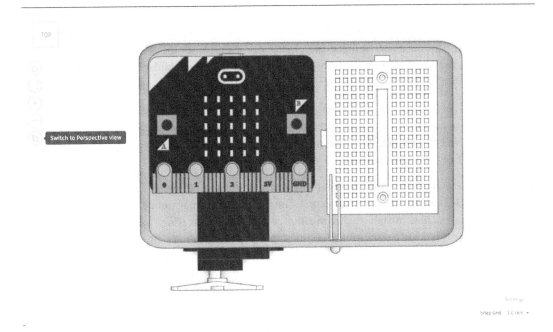

Figure 3.9: Looking at the top of a model in a 2D orthographic view

We're looking at the same model in *Figure 3.8* and *Figure 3.9*, but in *Figure 3.9*, we've enabled the orthographic view by pressing the **view** button on the right-hand side of the design window. This has flattened our 3D shapes, making the model look more realistic and representative from this viewing perspective. But as important as it is to switch to a 2D orthographic view when we're designing from a 2D perspective, it is equally as important to switch back to a 3D perspective view when we're looking at our models in 3D.

If not, our models will look skewed, as shown in *Figure 3.10*:

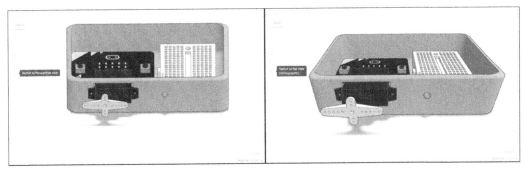

Figure 3.10: Looking at a design in the orthographic view (left) and the perspective view (right)

On the left of *Figure 3.10*, the orthographic view is still enabled, even though we are looking at our design from a 3D viewing angle. At first glance, it may be difficult to see the problem, but if we compare the left view to the one shown on the right, we can see that the model looks far more realistic in the image on the right, where the 3D perspective view is enabled and the orthographic view is not.

Zooming also acts differently, depending on the view we're working in. When in the default 3D perspective view, you will zoom in and out to wherever your mouse pointer is located when zooming with a scroll wheel on a mouse. But when working in a 2D perspective view, you will zoom in and out to the center of your design window, regardless of how you're zooming.

The key takeaway from this concept is that we need to understand the limitations that are presented when creating 3D models using a 2D screen. At times, the models may not be represented accurately, which can cause possible mistakes as we misinterpret what we're creating. You will find this to be an added challenge when you're designing different types of projects, such as the one we will be looking at later in *Chapter 16*. Fortunately, Tinkercad provides us with the tools to switch between a 3D perspective view and an orthographic view; we just need to know how and when to do it.

Summary

To elevate your Tinkercad design abilities, you need to increase your understanding and proficiency in utilizing the 3D design application. Being able to efficiently view your model, inspect its parts and features, and accurately interpret what it may look like after production is key.

In this chapter, we looked at scenarios where failing to change our views may be catastrophic, as well as the tools within Tinkercad to avoid making these common mistakes. Orbiting, panning, zooming, and changing between 3D and 2D perspectives are key tools that we will be applying in each following chapter.

When you're ready to apply these skills, flip the page to learn about strategies for designing through constructive solid geometry. Here, we'll see how just about anything can be created in Tinkercad!

4
Designing through Constructive Solid Geometry

In the final chapter of this part, we will be covering a key concept that encompasses the functionality and techniques for using Tinkercad: **constructive solid geometry**.

Prior to moving into chapters where we dive deeper into using specific tools, using 3D printers, or creating various projects from start to finish, this chapter is intended to ensure that each reader has a thorough and effective understanding of how to model in Tinkercad. To achieve this, this chapter is broken down into the following topics:

- What is constructive solid geometry?
- Working in groups
- Manipulating the shape parameters
- Hiding and locking shapes

Throughout this chapter, we will look at some of the common techniques and tools used in Tinkercad that are crucial for creating models successfully. Even more so, we will be looking at tips and strategies that are often missed by beginners, but that are crucial when we aspire to make complex designs and prototypes.

By the end of this chapter, you will be able to identify key strategies to create a design more efficiently as you become more comfortable with the hidden features and shortcuts in Tinkercad.

Technical requirements

We will be using Tinkercad as discussed in the *Technical requirements* sections of *Chapters 1-3*.

The example models shown in this chapter can be obtained through the following links:

- `https://www.tinkercad.com/things/0PtlZeDjTrn-bookshelf-example-from-chapter-4`

- `https://www.tinkercad.com/things/7gae9FdrelE-the-basics-of-csg-from-chapter-4`
- `https://www.tinkercad.com/things/5aZCLPFyHGh-coffee-mug-example`
- `https://www.tinkercad.com/things/1LVeapRNLZu-electronic-prototype-project`

What is constructive solid geometry?

As we've already learned, Tinkercad is one of many CAD programs available that allow you to create unique 3D models and designs. We've also learned that Tinkercad offers a more intuitive approach when compared to traditional CAD programs, because designing in Tinkercad feels familiar, sort of like playing with toy blocks as a child.

This concept of creation through blocks and shapes is a fundamental component of **constructive solid geometry**, or **CSG**. CSG is a 3D modeling technique that involves creating complex and unique shapes through combining simple or primitive ones. These shapes can be combined through **Boolean** operations, such as merging or cutting, to add or remove material from our final shape through a series of steps.

As an example, let's look at the model of a bookshelf designed in Tinkercad shown in *Figure 4.1*:

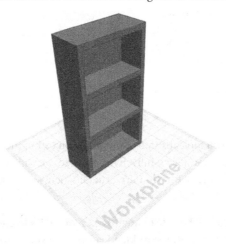

Figure 4.1: 3D model of a bookshelf designed in Tinkercad through the principles of CSG

Another key element of CSG is that there is rarely one way to create any model. A different combination of shapes and steps can often be employed to develop the desired result, which is another reason why Tinkercad is so intuitive and easy to learn; there's never one right answer! Looking at the model in *Figure 4.1*, how do you think it was created? How many shapes were used? Which ones were solid, and which were holes? Looking at *Figure 4.2*, we can see there are many possibilities to consider:

What is constructive solid geometry? 43

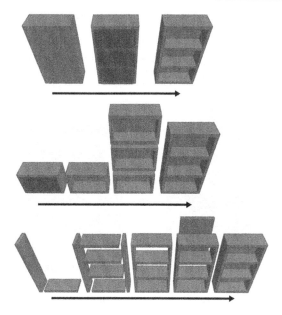

Figure 4.2: Different CSG techniques to create a model of a bookshelf in Tinkercad

Are any one of the approaches shown in *Figure 4.2* how you would have created this model, or did you think of another one? There is never one correct approach in CSG, but you may find that one approach requires fewer shapes or steps, which in turn often makes the modeling process simpler and more efficient. As we build our understanding of Tinkercad, we always want to strive for the most efficient approach to make modeling complex shapes a less daunting task.

When considering how Tinkercad is different than other CAD programs, we can look at which modeling principles they use. Many 3D modeling programs use **parametric** modeling principles instead of CSG, which offers a different approach to creating 3D shapes. Parametric modeling is a process of creating 3D designs through a series of sketches, profiles, and constraints. For example, the bookshelf model shown in *Figure 4.3* was created through a parametric approach:

Figure 4.3: 3D model of a bookshelf designed through the principles of parametric modeling

At first glance, the models shown in *Figure 4.3* and *Figure 4.1* look largely similar, although they were created through different approaches. The parametric modeling technique used to create the bookshelf in *Figure 4.3* starts with 2D profiles, or sketches, with constraints. Profiles are then turned into a 3D shape using an extrusion technique, as shown in *Figure 4.4*:

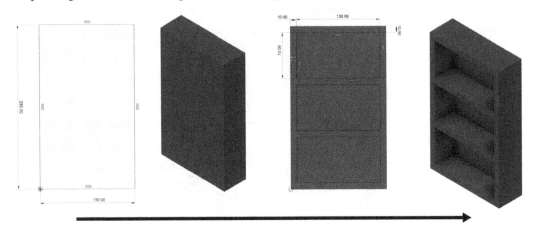

Figure 4.4: Steps taken to create a bookshelf through the principles of parametric modeling

Looking at *Figure 4.4*, we can see that two sketches were created and extruded to create the final product. Like CSG, there is rarely one approach to creating a model using parametric design principles, though the principles for this modeling technique are quite different.

So, which is better? In general, this really comes down to personal preference. Beginners often find CSG to be easier to learn as it takes a physical approach to creation, allowing you to select primitive shapes to create truly unique creations with less constraint than you might find in parametric design. In the professional world, designers often take advantage of multiple modeling techniques to create their desired designs through the most efficient approach.

While Tinkercad isn't a parametric design program, we can still apply constraints and dimensions, which we discuss more in *Chapter 6*. We can also sketch unique shape profiles using the Scribble tool as shown in *Figure 4.5*:

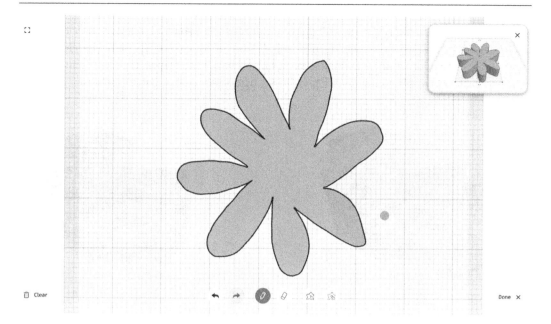

Figure 4.5: Drawing a unique shape profile using the Scribble tool

You can select **Scribble** from the Shape Panel by dragging it onto your **Workplane** as you would with any other shape. This will automatically open the Scribble tool window, which allows you to draw a unique outline or filled shape using the different drawing tools shown in *Figure 4.5*.

After clicking **Done**, your unique sketch will be made into a 3D model, which you can transform, grouped, and scale as you would with any other shape. If you want to edit the original 2D profile, select your shape and click **Edit Scribble** from the Shape Panel.

Modeling with shapes created using the Scribble tool in Tinkercad is still done through the principles of CSG, but it does offer a more versatile approach to creating unique shapes. As we consider methods for creating unique shapes or complex designs, we also must dive deeper into the CSG concept of working within groups.

Working in groups

As demonstrated in *Figure 4.2*, we can create unique models and designs in Tinkercad by combining different shapes through the principles of CSG. In this section, we will consider different strategies and techniques for using what might be the most important modeling technique in Tinkercad, **grouping shapes**.

As you already know, pressing **Group** will merge whatever shapes you have selected at the time, and **Group** cannot be pressed unless two or more shapes have been selected. If you have solid shapes, they will be merged into a single solid model, and if you have hole shapes, they will be merged by removing the hole shape from the solid shapes. This is demonstrated in the steps shown in *Figure 4.6*:

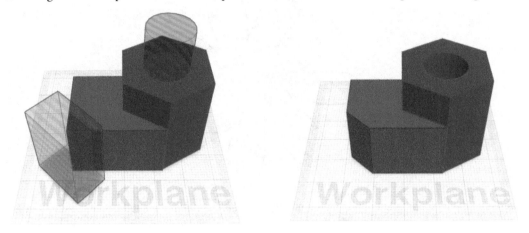

Figure 4.6: Four primitive shapes (left) are combined into a single solid shape (right) through grouping

To group your shapes together as shown in *Figure 4.6*, you can either press the **Group** button on the top toolbar of your screen or press *Ctrl + G* on your keyboard. You can also change the color of your group shapes from multicolor to a single color in the Shape Panel shown in *Figure 4.7*:

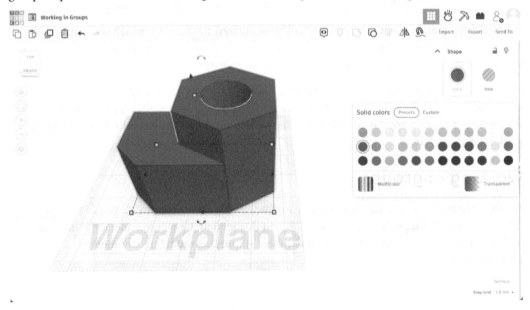

Figure 4.7: Changing the color of grouped shapes

If your solid shapes are different colors, Tinkercad may automatically make your grouped shapes multicolor by default, even if these shapes are combined into a single solid model. You can always customize the color and appearance of your shapes in your design using the Shape Panel, but it is important to note that we will still need to select the correct color material when 3D printing our models, as discussed later in *Chapter 10*.

When selecting multiple shapes to be grouped together, there are multiple techniques that can be used. You can hold *Shift* on your keyboard while clicking on each shape or press *Ctrl + A* on your keyboard to select all shapes if you want all the shapes in your design to be grouped. Alternatively, you can left-click and drag a selection box around your shapes in Tinkercad to select multiple shapes at the same time, as shown in *Figure 4.8*:

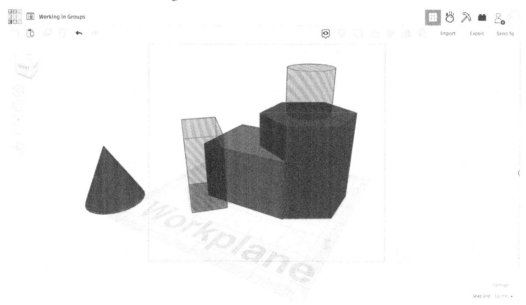

Figure 4.8: Creating a selection box around shapes in Tinkercad

Selection boxes allow you to select multiple shapes more easily without selecting everything in your design. But sometimes it's easy to select extra unwanted shapes using selection boxes, so you first want to change your view so that you can create your selection more accurately as shown in *Figure 4.8*.

It's also important to understand that the order in which you group objects will impact the result, as well as increase or decrease your modeling efficiency. For example, let's imagine that we are creating a model of a coffee mug as shown in *Figure 4.9*:

Figure 4.9: A 3D model of a coffee mug made in Tinkercad

How would you create this model? As we've learned, there is never one approach to modeling in Tinkercad, but we can certainly use grouping to create a model like the one shown in *Figure 4.9* more efficiently.

Method 1

If we were to create this model one shape at a time, for example, it might take us six steps as shown in *Figure 4.10*:

Figure 4.10: Steps taken for creating a 3D model of a coffee mug

Looking at *Figure 4.10*, the coffee mug was created using three grouping functions and four shapes. The steps are as follows:

1. Add a solid cylinder and a hole cylinder to the design.
2. Group the two cylinders to make a cup shape.
3. Add a solid torus shape and a hole box shape.
4. Group the torus shape to the hole box shape to cut it in half to make a handle.
5. Move the handle shape to align it with the cup shape.
6. Group the cup shape and handle shape to make the finished mug model.

However, we can take another approach to simplify the creation of this model instead.

Method 2

Alternatively, if we add the torus shape with our two cylinder shapes and group them all together at the same time, we could create this mug design in only two steps with one grouping function and three shapes as shown in *Figure 4.11*:

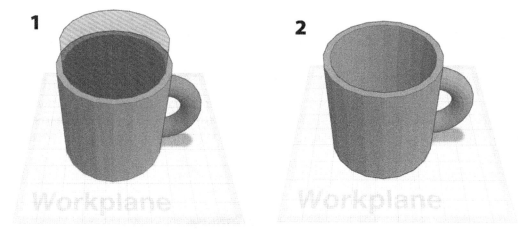

Figure 4.11: Modeling a coffee mug with fewer steps

The steps shown in *Figure 4.11* are as follows:

1. Add a solid cylinder shape, a hole cylinder shape, and solid torus shape to the design.
2. Group all three shapes together to create the finished mug model.

It is important that all three shapes in *Figure 4.11* are grouped together at the same time, or else the result would have been different. For example, see *Figure 4.12*:

Figure 4.12: Attempting to model a coffee mug in a different order

In *Figure 4.12*, the two cylinders were grouped first. Next, the torus was added and grouped to the cup shape. But because the hole cylinder was already merged in the first group, we are left with half of the torus in the mug, and half out. To fix this, we would need to add another cutting shape as shown in *Figure 4.10*, or group all the shapes at the same time as shown in *Figure 4.11*.

At first, it may not be clear what approach might be best to create your desired model. Fortunately, ungrouping your designs to make adjustments is very easy to do by either pressing the **Ungroup** button on the top toolbar, or pressing *Ctrl + Shift + G* on your keyboard. You can also make adjustments without ungrouping your shapes at all, as shown in *Figure 4.13*:

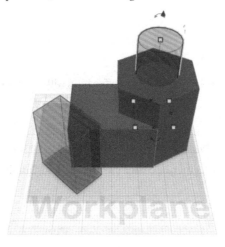

Figure 4.13: Making adjustments to grouped shapes without ungrouping them

By double-clicking on a grouped object, you will enter the group where adjustments can be made to the primitive shapes used. Looking at *Figure 4.13*, we can tell that we are within a group because a red shadow box appears around our grouped shapes. After adjusting the primitive shapes within the group, we can click anywhere on our screen to exit the group.

As for manipulating and adjusting your basic shapes, there is more than one approach that you can utilize. In addition to the fundamental approach of dragging the transformation handles shown around selected shapes to change their size or position, we can also utilize the shape parameters for greater precision.

Manipulating shape parameters

Whenever you select a shape, the Shape Panel opens on the right side of your screen where you can find options to make the shape a solid or a hole, as well as change the color.

Below these options are what we call **shape parameters**, which allow us to manipulate our shapes with more precision and versatility. As you already know, you can drag the corners of a shape to change its size, but you can sometimes also find parameters to change the size within the Shape Panel as shown in *Figure 4.14*:

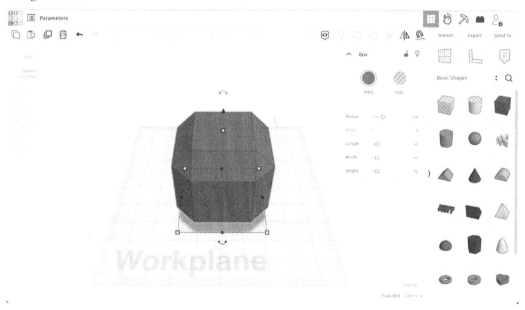

Figure 4.14: Adjusting the shape parameters for box shapes

To change the size of a box, you could drag the slider for whichever parameter you want to adjust within the Shape Panel, or you can enter a specific size instead. The size you enter will be in whichever unit you are working with (millimeters, inches, blocks).

But the Shape Panel also often offers parameters that cannot be adjust elsewhere, such as adding a corner radius to the box shown in *Figure 4.14*. By adjusting the **Radius** shape parameter, we can add a rounded edge to our box shape in addition to adjusting the **Steps** value, which changes the overall smoothness of the sides.

It's also important to understand that the shape parameters will vary from shape to shape. Take the cylinder shape parameters shown in *Figure 4.15* for example:

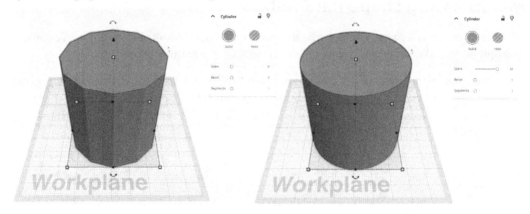

Figure 4.15: Adjusting the shape parameters for a cylinder shape

As you can see in *Figure 4.15*, the cylinder shape parameters do not include length, width, height, or radius like the box shape does. Instead, we see a slider to adjust the **Bevel** value, which allows us to put a beveled edge on the shape, as well as adjust the number of **Sides**.

The cylinder shown to the left has few sides, making it appear more blocky and polygonal, while the **Sides** parameter was increased for the cylinder shown on the right, making it appear smoother. Not only does adjusting the shape parameters change the appearance of the shapes in our 3D designs, but these changes will also be present if we choose to 3D print these models as discussed later in this book.

Every basic shape in Tinkercad has unique parameters that allow you to adjust the shape's size or appearance in greater detail than you can just by dragging it within the design window. Play around with each shape to get comfortable with the parameters and familiarize yourself with what can be manipulated! As you add and manipulate shapes, you might find that your workspace starts to become crowded. In these circumstances, you can take advantage of Tinkercad's tools for hiding or locking shapes.

Hiding and locking shapes

Earlier in this chapter, we discussed that it can sometimes be difficult to select just a few shapes in your design without changing your view around. Another strategy to assist in working with complex design windows with a lot of shapes is to hide the shapes when they are not needed, as shown in *Figure 4.16*:

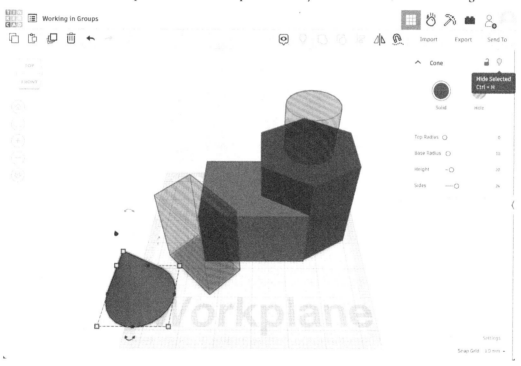

Figure 4.16: Using the Shape Panel to hide selected shapes

To hide a shape, you must first select it so that the Shape Panel appears. At the top of the Shape Panel, you can click on the **Hide Selected** button, which looks like a lightbulb, to hide the selected shape. You can select and hide more than one shape at a time, and you can also press *Ctrl + H* on your keyboard to hide selected shapes instead of using the button in the Shape Panel.

Hiding shapes does not delete them from your design, but instead removes them from your view. This prevents you from accidentally selecting shapes and allows you to clean up your workspace by hiding things that are not needed. Unhiding a shape is also easy to do, as shown in *Figure 4.17*:

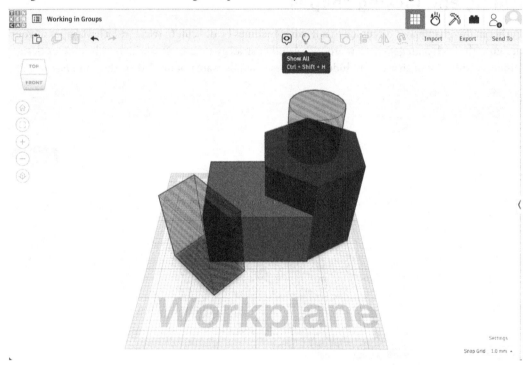

Figure 4:17: Pressing the Show All button to show hidden shapes in the design

The **Show All** button also looks like a lightbulb and is located on the top toolbar in the design window. This button can only be clicked if there are hidden shapes in your design, otherwise it will remain grayed out. You can also press *Ctrl + Shift + H* on your keyboard to show all hidden shapes as well.

Another strategy for designing complex models is locking shapes so that they cannot be manipulated. Unlike hiding shapes, locked shapes are still visible in your design. But like hidden shapes, locked shapes cannot be accidentally moved, grouped, or selected. To lock a shape, you must first select it for the Shape Panel to appear as shown in *Figure 4.18*:

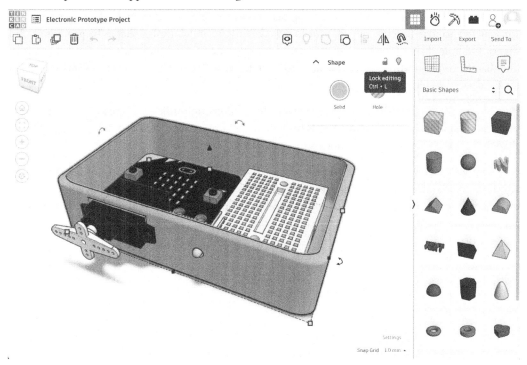

Figure 4.18: Locking and unlocking selected shapes using the Shape Panel

You can toggle the **Lock Editing** button shown in the Shape Panel to lock or unlock selected shapes. Alternatively, you can press *Ctrl+L* on your keyboard to lock and unlock selected shapes as well. After locking a shape, it will be highlighted in purple to show that it is locked. You will still be able to click on locked shapes, but you cannot move, scale, delete, or group them without unlocking them first.

As you strive to create more complex designs with many different shapes, consider hiding shapes that are not needed or locking shapes that must stay in place or remain of the same scale to make modeling easier and more efficient.

Summary

Let's look back at what was covered in this chapter to consider the strategies and techniques identified prior to moving on to the next part of this book.

Tinkercad offers a unique and intuitive approach to designing 3D models through the principles of constructive solid geometry. In CSG, you can make unique and complex shapes by grouping solid shapes and hole shapes together. There is never one approach to designing any model, but you may find that the order in which shapes are grouped can decrease the number of steps and shapes needed for the desired result.

We can also manipulate shapes to enhance our designs through unique shape parameters, which vary from shape to shape. As your designs become more complex and crowded, you can hide or lock shapes to make designing unique models more efficient.

Now that we've covered strategies for successful 3D modeling, we are ready to move on to using advanced tools and features to enhance our designs. We will continue to put the concepts learned in this chapter to the test as we use CSG to make real-world projects and designs. When you're ready to get started, flip the page to start part two of this book!

Part 2: Advanced Tools and Features to Enhance our Designs

Building on the key strategies and topics introduced in *Part 1*, this part of the book will allow us to dive deeper into the advanced modeling techniques that are possible using Tinkercad. Each chapter within *Part 2* focuses on specific tools or features that you may want to incorporate into your projects. As we cover these topics, we will work to push the envelope past basic modeling techniques by identifying how Tinkercad can be used to make models that are aesthetically pleasing, efficiently designed, and uniquely your own.

This part includes the following chapters:

- *Chapter 5, Creating and Manipulating Text Features*
- *Chapter 6, Using the Ruler and Workplane Tool to Dimension Our Designs*
- *Chapter 7, Tools to Manipulate and Pattern Multi-Part Designs*
- *Chapter 8, Importing Models and Designs*
- *Chapter 9, Making Our Own Shapes*

5
Creating and Manipulating Text Features

Throughout the first section of this book, we looked at key strategies for successful 3D modeling, including what is possible through the Tinkercad design application, how to brainstorm and prepare your designs for 3D modeling, and tools that may assist you in creating a successful design workspace.

As we begin the second part of this book, we will look to apply the strategies discussed and demonstrated in earlier chapters as we dive deeper into using specific tools and techniques for advanced modeling. In this chapter, we look at the different ways to create text features in our designs through the following topics:

- Typing and grouping text
- Using individual characters
- Using text shape generators
- Thinking outside the text box

As you progress through this chapter, you will find that there is more than one way to create a text feature in Tinkercad. And as a designer, it is important to learn different tips and tricks so that you may enhance your designs beyond the basics of making a two-dimensional keychain.

It is also important to begin to consider how we can design our models so that they will be able to be manufactured more successfully using technology such as 3D printers. We will be looking at these concepts in greater detail later in *Chapter 10*, but we will also begin to consider these concepts now as we make unique text features in our designs.

By the end of this chapter, you will be able to determine which approach to creating text features may suit the needs of your design best, as well as to consider how text shapes can become a focal point in your 3D printed designs.

Technical requirements

We will be using Tinkercad as discussed in the *Technical requirements* sections of *Chapters 1-3*.

An editable model of the complex text example shown in this chapter can be accessed on Tinkercad at `https://www.tinkercad.com/things/4KB8p03Sjqe-text-example-from-chapter-5`.

Typing and grouping text

In Tinkercad, the **TEXT** shape can be found under the **Basic Shapes** category and selected like any other shape. After dragging **TEXT** onto our *Workplane*, we can see that the shape parameters differ from typical shapes as shown in *Figure 5.1*:

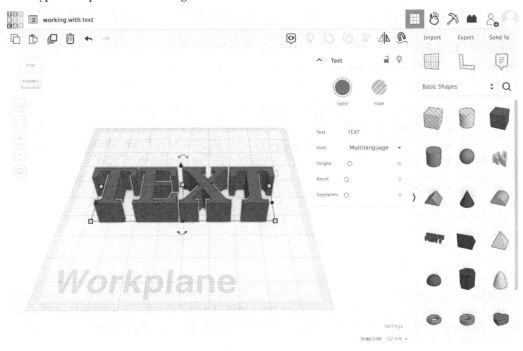

Figure 5.1: Adjusting the basic text shape parameters

We can enter a single line of text, as well as choose between four different fonts that offer simple serif and sans-serif appearances. We can also adjust the height of our text, though the text shape can be dragged and scaled like any other shape we can work with.

To add a **Bevel** to the face of your text, you should also increase the **Segments** parameter as shown in *Figure 5.2*:

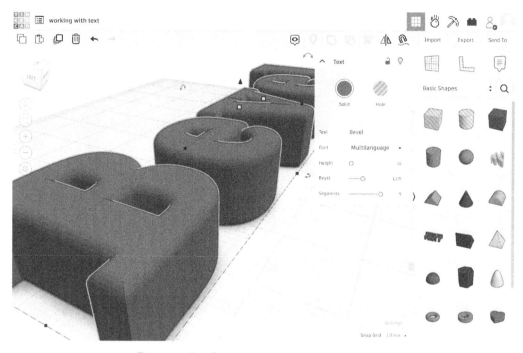

Figure 5.2: Beveling text using the shape parameters

If **Segments** is not increased, **Bevel** just adds an outline around your text that merges the shapes together. When both the **Bevel** and **Segments** parameters are increased as shown in *Figure 5.2*, a slanted edge becomes more apparent on the front and back faces of the text shape.

As we design parts for production in Tinkercad, text can be used to personalize or enhance our designs. Some ways to do this are the following:

- Extrusions
- Pockets
- Orientation

Let's look at each of these in the subsequent sections.

Extrusions

The most common way to group text to an existing shape is through an **extrusion-type method** as shown in *Figure 5.3*:

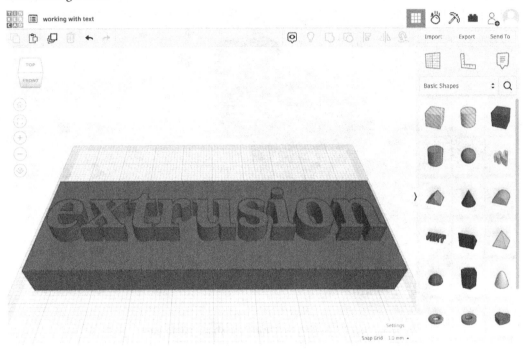

Figure 5.3: Grouping text so it extrudes from a shape

As shown in *Figure 5.3*, the extrusion-type method raises the text feature out of a part. This offers a very noticeable presence in your design digitally, but it may also create challenges when 3D printing. We want to make sure the text is touching the other shapes rather than floating above it as shown in *Chapter 3*. And extruded text typically works best when the extrusion is vertical as shown in *Figure 5.3* so that the text can build upon itself as the part is printed.

If the part was rotated so the extrusion hangs outward, as shown in *Figure 5.4*, challenges may arise.

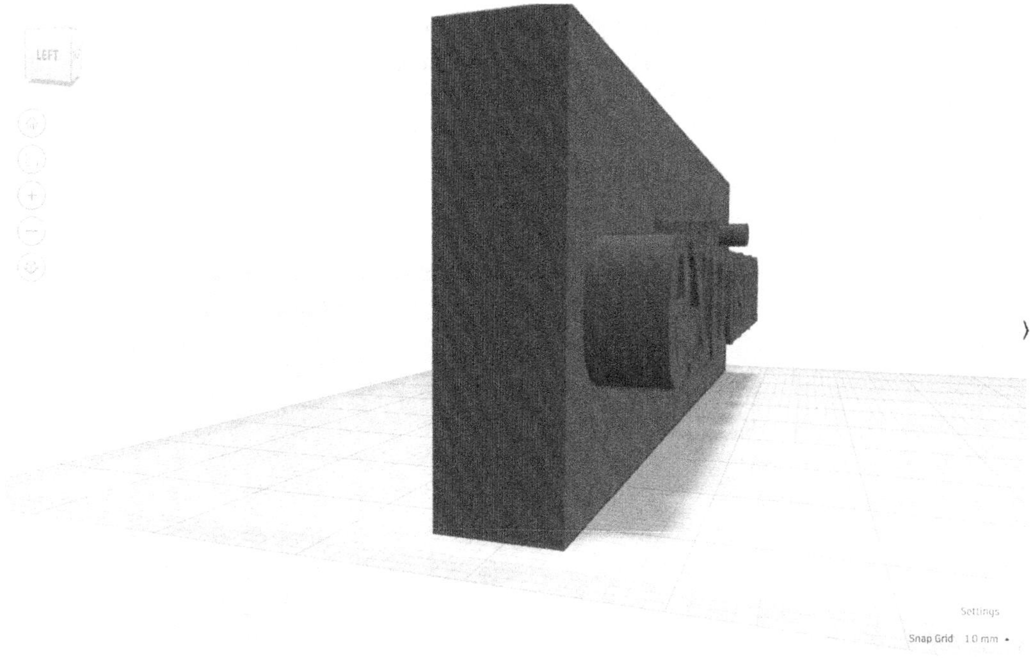

Figure 5.4: Extruded text shown as an overhang

The text in *Figure 5.4* is modeled as an **Overhang**, a part with nothing to support it as it is 3D printed. The challenge with this is that the model may "droop" during the 3D printing process as gravity takes over.

There are some ways around this, which will be discussed later in *Chapter 11* as we get into designing specifically for 3D printing, but consider keeping text extrusions vertical or with minimal overhangs if possible. Alternatively, you may also consider modeling your text so that it is a pocket instead of an extrusion.

Pockets

The term for this process may vary depending on where in the CAD world you are coming from, but when we turn our text into a *hole* shape so that it cuts into our model instead of extruding from it, I consider that to be a **pocket**. The result of this method is shown in *Figure 5.5*:

Figure 5.5: Using text as a hole shape to create a pocket in a design

To create a pocket like the one in *Figure 5.5*, you must first select and adjust your text shape as desired. After changing it to a hole, you can then group it with another shape to create the pocket. Like extrusions, we don't want to make our pockets cut too deep as that may cause them to cave in when 3D printed vertically. But pockets are typically a bit more rigid in their design, which allows for greater detail and a higher quality finish.

Both pockets and extrusions can be strengthened by adding a **Bevel** to the text face as shown earlier in this chapter. **Bevel** removes the hard edge of the text, which makes 3D printing letters a bit easier to do as overhangs will be reduced. Typically, hard edges are the first thing to fail when 3D printing quickly or on a lower resolution 3D printer, so it's a good common practice to enable bevels when adding a text feature to our design with the intention of production.

Orientation

The last thing to consider when modeling with text is the **text orientation**. Let's say we want to put a text pocket on the bottom of a vase we plan to 3D print, as shown in *Figure 5.6*:

Typing and grouping text | 65

Figure 5.6: Modeling a text pocket on the bottom of a transparent vase from a top view

In *Figure 5.6*, the text shape is set to be a hole while the vase is set to be transparent so we can see through it. From this top view, all looks well. However, what would happen after we 3D print this vase and look at the bottom of our model?

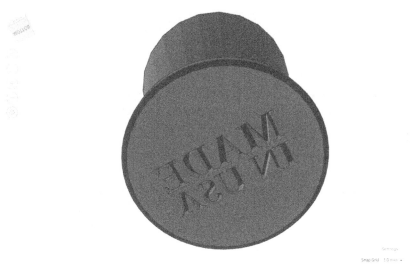

Figure 5.7: Looking at the bottom of the vase model shown in Figure 5.6

Figure 5.7 shows the same vase model shown in *Figure 5.6*, but from the bottom after the shapes have been grouped. From this perspective, we can see that our text is backward.

As you are modeling, it is important to change your views so that you can see how your model appears from different perspectives, as discussed in *Chapter 3*. To model this vase more effectively, we should design the text feature from the bottom view as shown in *Figure 5.7*.

If we do this, we can use the **Mirror** tool to resolve this issue as shown in *Figure 5.8*:

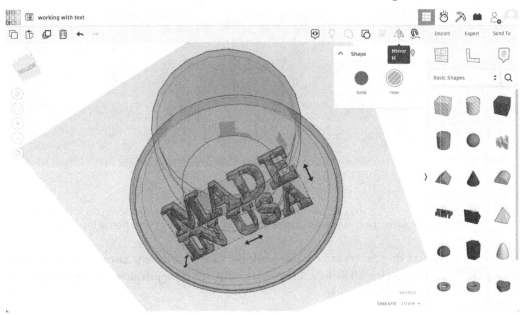

Figure 5.8: Mirroring text

After selecting a shape, the **Mirror** tool can be selected from the top toolbar to reflect the text shape across a plane. You can also press *M* on your keyboard after selecting a shape to mirror it. With the text shape mirrored as shown in *Figure 5.8*, it may look wrong from a top view, but the finished product would be correct when looking at the vase model from the bottom.

When creating multiple lines of text or different size words, there are a few different approaches that can be used. I often find that making multiple text shapes is usually the easiest and most effective. However, you may find that working with individual characters may be the best approach for your design.

Using individual characters

At times, you may find that using the *TEXT* shape is not the most effective strategy for creating a text feature. Fortunately, there are a few different methods for creating and working with text.

If you want to work with individual characters for example, you could use the text shape to type just one letter or number. Alternatively, you can also find individual characters under the **Design Starters** shape collection as shown in *Figure 5.9*:

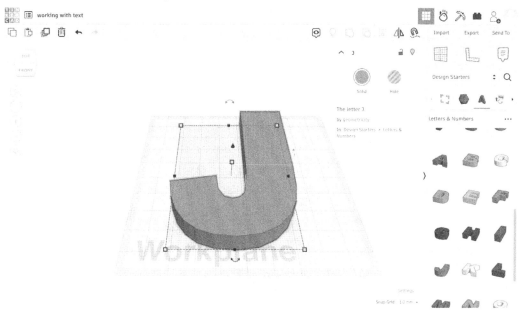

Figure 5.9: Selecting individual letter and number shapes in Tinkercad's design starters category

After choosing the **Design Starters** shape collection in your Shape Library, select the **Letters and Numbers** section or browse by searching for a specific shape. Here you can select individual characters and drag them into your design. There are no parameters to manipulate, however, as these individual characters are primitive shapes that can only be adjusted in scale and color as shown in *Figure 5.9*.

In addition to the design starters, there are more methods for working with text. Another shape collection to look at is the shape generators.

Using text shape generators

Shape generators can be found in their own category within the Tinkercad Shape Library. Later in *Chapter 9*, we dive deeper into the workings of shape generators, but here we will look specifically at some of the generators designed for working with text.

There are dozens of shape generators that vary in design, so you may find that searching is easier than scrolling through the different pages. The first text shape generator we will look at is called **script**, as shown in *Figure 5.10*:

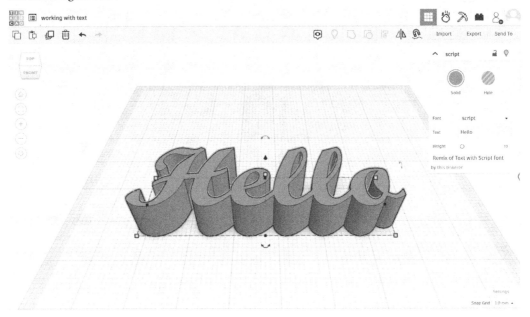

Figure 5.10: Using the script shape generator

The **script** shape generator is very similar to the default *TEXT* shape, though we lose the ability to add a **Bevel** to the face of the text. However, additional fonts are added to the shape parameters, which makes creating text a bit more unique and exciting.

There's another shape generator that works similarly to the standard *TEXT* shape and *script* text generator, but with a different outcome as shown in *Figure 5.11*:

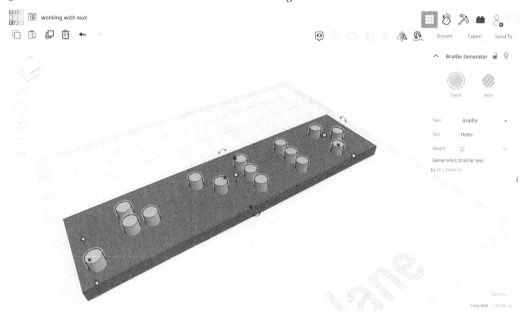

Figure 5.11: Using the Braille shape generator in Tinkercad

The **Braille Generator** allows you to type text and adjust the **Height** in the shape parameters, and the model is then created in either **Braille** or **Hang Talk** based on your settings in the shape parameter window.

You can scale this text feature like any other shape, but it is important to keep the scale proportional so that the Braille characters are not distorted. Hold *Shift* on your keyboard as you drag and change the size of this text feature and use the shape parameters to adjust their height as shown in *Figure 5.11*.

Another text shape generator to look for is called **text ring**, and as the name suggests it allows us to wrap text around a ring as shown in *Figure 5.12*:

Figure 5.12: Using the text ring shape generator

Like the other text shapes, you can enter your text in the parameters window as well as switch between a few different preset fonts. To effectively use the **text ring** generator, you may find that both the parameters for **Text Gap** and **Thickness** need to be adjusted in unison with an overall scale to get the text to fit around cylindrical shapes as shown in *Figure 5.12*.

The final text shape generator we are looking at is called **curved words**, which offers a different approach to manipulating text features than what we've covered thus far. Like all shape generators, the curved words generator was designed by a member of the Tinkercad community, and its functionality and performance may differ from the basic shapes created specifically by the Tinkercad team. To start, select the **curved words** shape and move it onto your **Workplane** as shown in *Figure 5.13*:

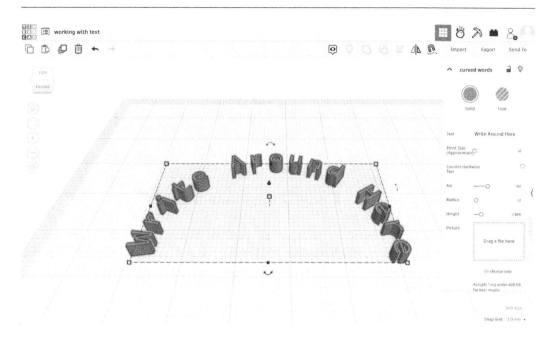

Figure 5.13: Using the curved words shape generator

This shape generator does not have any alternative fonts to choose from, but it does allow us to adjust the arch our text is written along in a horizontal plane. You can adjust the **Radius** and direction of the arch, as well as the length. By combining these parameters with the standard scaling tools, you can use this shape generator to create text extrusions or pockets on a flat surface of your design.

This is not the only way to arch text around a curve, however. Later in *Chapter 7*, we look at creating patterns with our shapes that can produce a similar result.

Thinking outside the text box

In the final section of this chapter, we will be applying some of the different techniques we covered as we consider how text shapes may be used outside of the traditional methods. For example, let's say you have a unique shape with asymmetric curves and edges as shown in *Figure 5.14*:

Figure 5.14: An example shape shown in Tinkercad

How could we get our text to wrap along the face of this part? We could use individual characters and space them out along the surface, which would work. But that method is certainly tedious. And if we wanted our text to wrap or fit along a shape even more complicated than the one shown in *Figure 5.14*, then this might not be feasible.

Instead, we can use the concepts of CSG to create custom profiles using our text shapes, which can wrap around any surface. To do so, follow these steps:

1. Start by creating a text feature using the standard '**TEXT**' shape, or by using one of the flat generators we covered in this chapter, as shown in *Figure 5.15*:

Figure 5.15: Creating a text feature in front of the example shape

It's important to scale the text feature so it fits in our example shape as desired, but we aren't connecting the text to the example shape just yet. We will be able to adjust the scale proportionally later, but the wording, spacing, and fit for the text should all be adjusted at this stage.

2. Increase the depth of the TEXT shape so that it is greater than the shape you want to wrap the text around. We are keeping the text separate from the example shape, but we want to set it to **Hole** instead of **Solid** using the shape parameters window shown in *Figure 5.16*.

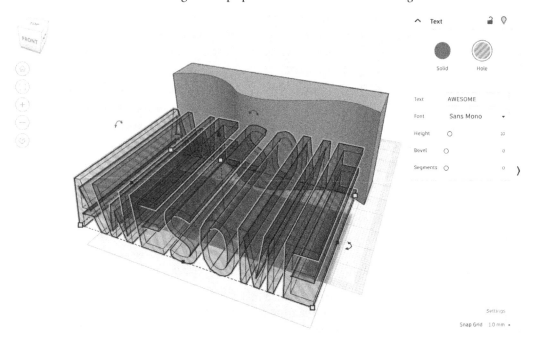

Figure 5.16: Adjusting the depth and properties of the text feature

3. Next, we want to insert a **Box** shape that is wider and taller than the **TEXT** shape, but not as deep, as shown in *Figure 5.17*:

Figure 5.17: Inserting a box around the text feature

This **Box** shape does not need to be aligned in any particular way except that the text feature (hole) should extrude from both the front and back sides of the **Box** so that it passes all the way through the shape. Ensure that the box shape is bigger than the original example shape we want to wrap around, as shown in *Figure 5.17*. Once the size is adjusted, **Group** the text feature to the **Box** shape so that the **Text** cuts a hole straight through the box.

We will use this text/box shape as sort of a CAD mold or die cutter, if you will. This will allow us to cut a text shape out of our example shape that perfectly matches the unique geometry of this shape as it is made from it. For the next step, make a copy of the example shape and move it off to the side as shown in *Figure 5.18*:

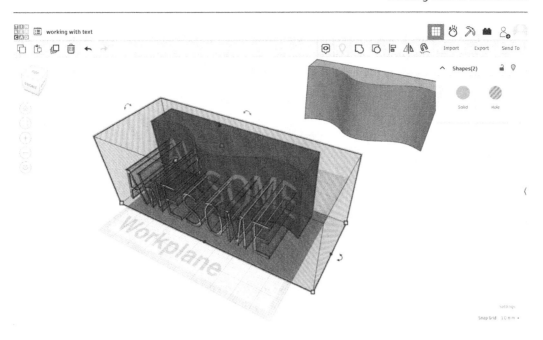

Figure 5.18: Copying the example shape and adjusting the cutting shape

4. Once a copy of the example shape has been made, make the box text shape a **Hole** and position it around one of the example shapes. Ensure that the entirety of the example shape fits within the box as shown in *Figure 5.18*. After aligning the two shapes, **Group** them together to cut your text as shown in *Figure 5.19*:

Figure 5.19: The result of grouping the box and example shapes

As shown in *Figure 5.19*, our once flat text feature is now a 3D model that follows the curves of the original example shape shown in *Figure 5.14*. This was created through the holes of the box shape we made and grouped together with the example shape through the fundamental principles of CSG. This unique creation might be the entirety of your design in itself, or we can take it one step further and group it to the example shape, as shown in the next step.

5. Bring the example shape we copied and moved off to the side back onto your Workplane and align it with the curved text feature we created in *Figure 5.19*. As shown in *Figure 5.20*, the text wraps perfectly along the surface of the example shape as it was cut from its copy.

Figure 5.20: Grouping the curved text shape with the copied example shape

It's impossible for me to present every possibility or creation that you can do in Tinkercad, but consider that this example shape doesn't need to be a curvy block. Instead, it could have been a silhouette of an animal, the skyline of a city, or even the globe. Likewise, the tools and strategies we used with the text features could have been done with another shape or design as well, which could have created an entirely different model. The key takeaways are the steps and tools used, as well as how these tools can be combined in countless variations to make truly limitless designs.

Summary

Looking back at the topics covered in this chapter, we can see that text may not only be a feature of your models, but a model in and of itself. Tinkercad offers a wide range of tools to create and manipulate text features in our designs starting from the basic text shape.

But if you want to create text out of a standard box, you can work with individual characters or even choose from a range of shape generators instead. There is no one approach that works best in every circumstance. As a designer, it is important to identify what tools you have available so that you can choose the best approach when creating your models.

By applying previously learned concepts for designing through CSG, we can create truly unique models that rely on text for their functionality and performance. We will revisit text tools later in this book as we learn more tools and strategies for creating complex designs.

When you're ready to learn how to apply greater precision by adding dimensions to your design, turn the page to learn more about the advanced measurement tools available in Tinkercad.

6
Using the Ruler and Workplane Tool to Dimension Our Designs

Earlier in this book, we learned how important measurements, or **dimensions**, are as we create designs modeled for production. Even if we are designing something as simple as a keychain, it is crucial for the keychain to not be so large that it does not fit in our pocket, or so small that it isn't even big enough to be 3D printed.

Despite its simplicity and intuitive nature, Tinkercad allows complex dimensions to be applied to our shapes and provides tools that increase the accuracy of our designs. We will investigate and learn how to use these tools through the following topics:

- Working with the grid
- Using the ruler tool
- Using the workplane tool

At times, you may be using Tinkercad to create complex worlds, models, or scenes that may never leave the screen of your device. In these cases, dimensions may not be as important as a sense of scale and proportions. But as we progress through this book and learn about tools that may guide us through creating successful models for 3D printing and production, the skills gained in this chapter cannot be underestimated.

By the end of this chapter, you will not only be able to determine the size of the parts within your design, but also set dimensions between them, design according to constraints, and understand the tools available to make designing accurate models more efficient in Tinkercad.

Technical requirements

We will be using Tinkercad as discussed in the technical requirements sections of *Chapters 1-3*. The example model shown in this chapter can be obtained from: https://www.tinkercad.com/things/1GZwNIE1KSL-workplanes-and-dimensions-example

Working with the grid

In every Tinkercad design window, there is a **Workplane**, which looks like a grid, as shown in *Figure 6.1*:

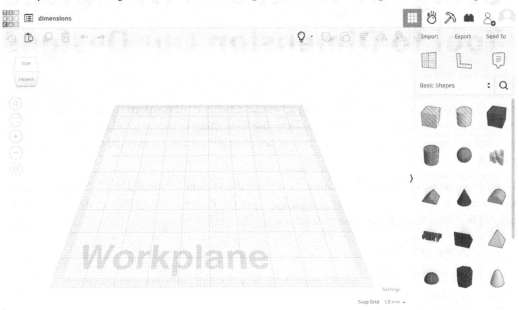

Figure 6.1: The default workplane in a Tinkercad 3D design

This workplane is the starting point of our design. Every shape we drag out will be placed onto it, and we can of course move and adjust our designs all around this grid. But the grid can be adjusted to better suit the needs of your design, as shown in *Figure 6.2*:

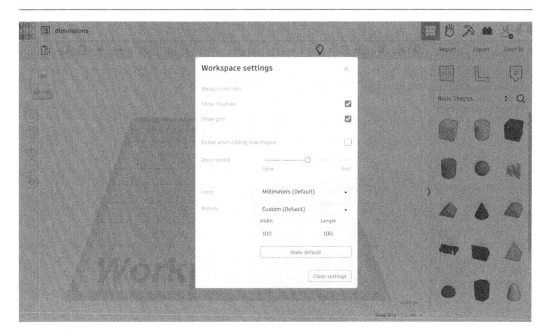

Figure 6.2: Adjusting the Workspace settings

By pressing **Settings** in the bottom-right corner of our design window, we are shown the **Workspace settings** menu. Here, you can choose to show or hide the grid (**Show grid**), as well as adjust the background color (**Background color**) and appearance of shadows (**Show shadows**) in your model.

You will see that the default size of your workplane is 100 by 100 millimeters. This can be increased up to 1,000 by 1,000 millimeters or made smaller. If I plan to 3D print my design, I like to adjust the size of my workplane to match the size of my 3D printer's bed to ensure that my design will fit during production, as discussed later in *Chapter 11*. Tinkercad also provides some preset 3D printer bed sizes to choose from in the **Presets** menu. It's important to note that your design can be off the workplane and even larger than 1,000 mm x 1,000 mm, so the workplane is there more as a guide rather than a constraint. If desired, you can also change your unit between millimeters, inches, or blocks, and save your settings as the default for all future 3D designs you create.

To dimension a shape in Tinkercad, you can drag the corner handles to change its size. If you click on a corner handle rather than drag it, you will find that the dimensions appear and stay on the screen momentarily. When shown, you can click on a dimension and type a desired size for whichever unit you are working in, as shown in *Figure 6.3*:

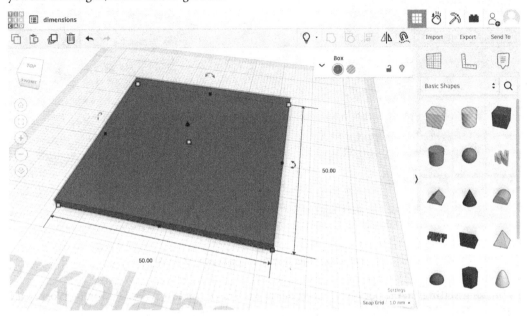

Figure 6.3: Dimensioning a shape on the workplane in Tinkercad

Your dimension will be set when you press *Enter* on your keyboard, or when you simply click away from the shape. You can also press *Tab* to switch between the dimension boxes for your shape as you apply measurements for the width (X), length (Y), or height (Z) of your shapes. If you're dimensioning something like a cylinder and you want to adjust its diameter, you would dimension both the length and width to be the same.

You can also dimension the elevation of a shape in relation to the workplane, as shown in *Figure 6.4*:

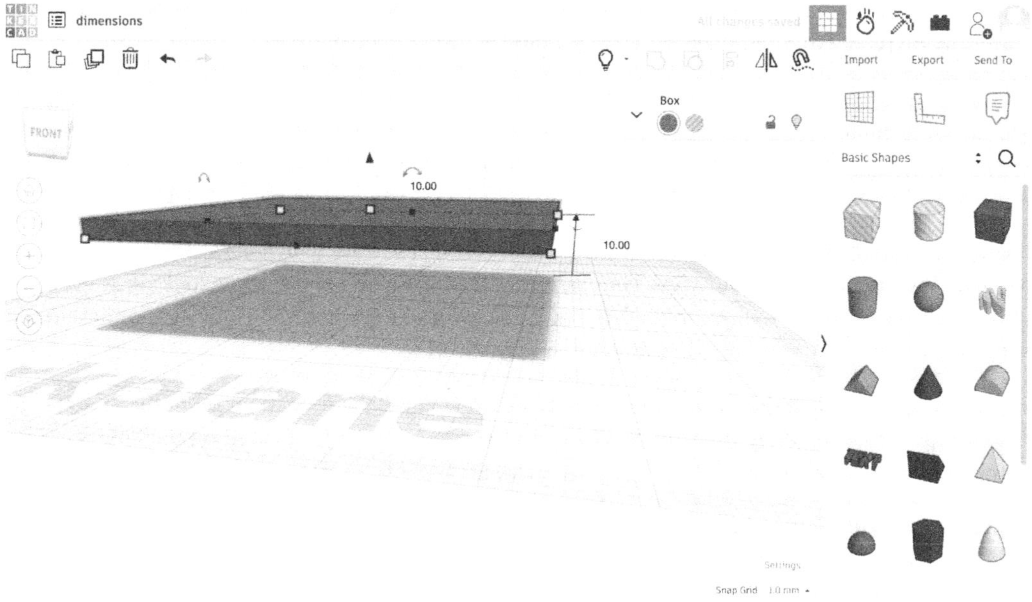

Figure 6.4: Setting a dimension between the bottom of a shape and the workplane

After raising a shape off the workplane using the black cone-shaped handle, a dimension box, like the one shown to the right side of the shape in *Figure 6.4*, will appear, which shows the distance between the bottom of your shape and the workplane surface. You can then click on this dimension box to set an accurate elevation for your shapes.

As you drag the corner handles of a shape to change its size, you will find that the dimensions snap or jump rather than increase gradually. This snap increment is set by the **Snap Grid** menu, which is in the bottom-right corner of your design window, as shown in *Figure 6.5*:

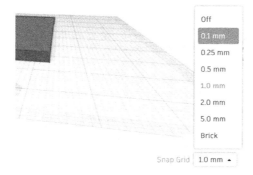

Figure 6.5: Adjusting the workplane Snap Grid units

If you have a smaller **Snap Grid** set, you will find that the dimension of your shape increases more precisely as you drag its corner handles to change its size. Likewise, the bigger the **Snap Grid**, the bigger the increase every time you drag your mouse across the screen.

While typing in a dimension may be the most accurate way to change the size of a shape, as demonstrated in *Figure 6.3*, **Snap Grid** really comes in handy in the movement of your shapes or changing the size in tight spaces, as shown in *Chapter 16*. You will also find that the most accurate way to move a shape is by using the arrow keys on your keyboard using the increments set using **Snap Grid**.

If you press the arrow keys on your keyboard, your shapes will move incrementally across the workplane based on the orientation of your view. For example, pressing the left arrow when looking at the front of your design will make it move to the left, but if you are looking at the right side of your design the left arrow will make it move forward in relation to the workplane.

The distance your shape moves for each press of an arrow key is the increment set in the **Snap Grid**. For example, if I wanted my shape to move exactly 10 mm to the right. I could set the **Snap Grid** to 5 mm, then press my right arrow key twice when looking at it from the front view.

It's also important to note that sometimes entering accurate dimensions or using the **Snap Grid** to size our projects is not always necessary. But these tools can still come in handy to set overall scale and proportions between different parts. Sometimes, when 3D printing projects, the overall size is more important than the individual pieces, as discussed later in *Chapter 15*.

But when measurements are crucial, there is another tool that can be used when setting distances and dimensions between multiple shapes in our design called the ruler tool.

Using the ruler tool

The **ruler tool** is not something we typically have on all the time in our 3D designs, but it can be enabled by dragging the ruler from the top of our shapes panel onto our workplane, as shown in *Figure 6.6*:

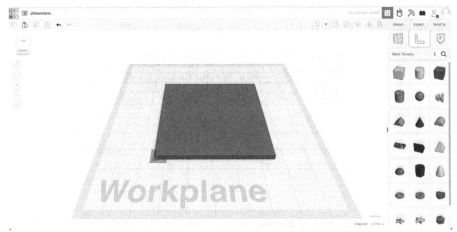

Figure 6.6: Enabling the ruler tool in a 3D design

Figure 6.6 shows the process of dragging the **Ruler** tool from the top of the shapes panel and dropping it onto the **Workplane**. After selecting a shape with the ruler tool enabled, all dimensions will be shown as long as the shape remains selected, as shown in *Figure 6.7*. The ruler tool also shows the distance between selected shapes and the origin of our ruler, which can be repositioned by dragging the ruler's *origin point*. The origin of the ruler is in the corner where the two axes meet and is represented by a dot, as shown in *Figure 6.7*. If you would like to flip the axis of your ruler so it faces a different direction, simply click on the *origin point* to rotate the ruler tool.

As an example, let's consider that we are placing 5 mm holes along the top of the plate-like shape shown thus far in this chapter. After designing the shapes and placing them in an approximate location, we can drag the ruler so that its origin position is set to be aligned with the bottom left corner of our plate shape, as shown in *Figure 6.7*:

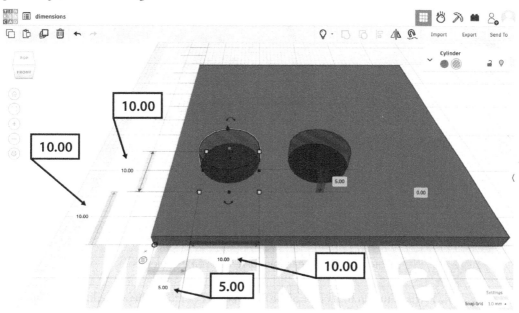

Figure 6.7: Positioning the ruler tool to dimension multiple shapes

As I click on each hole shape, the dimensions between the hole shape and the left and front edges appear in green. If I want to set the holes to be 5 mm from the left edge, 10 mm from the front edge, and 5 mm apart, I can use the ruler to show and set each of these dimensions in relation to the corner of my plate shape, as shown in *Figure 6.7*.

You will notice that the distances shown in *Figure 6.7* are based on the edge, or **endpoint**, of my hole shape. So, if my holes are 10 mm in diameter and I want them to be 5 mm apart, the distance from the second hole to the left edge of my plate is actually 20 mm (the first gap of 5 mm + the first hole diameter of 10 mm + the second gap of 5 mm). In this scenario, I would prefer to dimension my holes from their center positions, or **midpoint**. This can be set using the ruler tool, as shown in *Figure 6.8* and *Figure 6.9*:

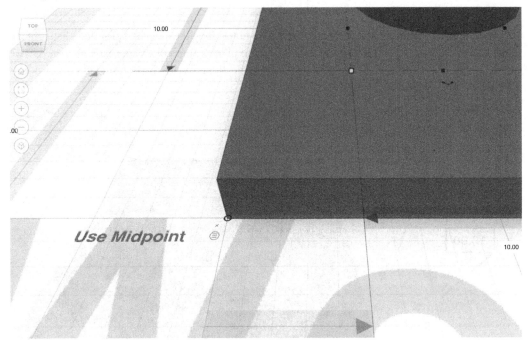

Figure 6.8: Toggling to midpoint mode on the ruler tool

By pressing the toggle button in the corner of the ruler tool, we can toggle between dimensioning from the endpoint or midpoint of our shapes, as shown in *Figure 6.8*. When the toggle button displays **Use Midpoint**, that means you are currently measuring based on endpoints, and vice versa.

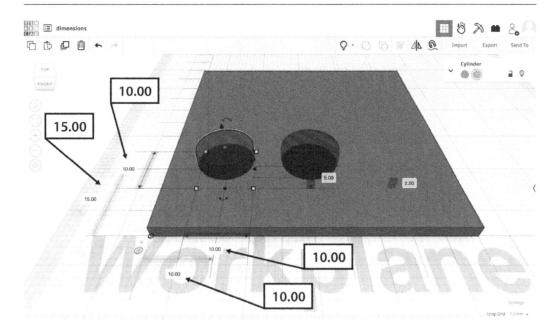

Figure 6.9: Showing dimensions between shapes based on midpoints

The position of the shapes does not change when we toggle between these modes, only the point at which we are measuring changes, as shown in *Figure 6.9*, which is now measuring this shape from the midpoints rather than the endpoints. You might also find that it is easier to drag and move the ruler's origin as you are dimensioning between shapes rather than leaving it in one fixed position. Being able to toggle between these different measuring modes is not always needed, but sometimes comes in handy when you're aligning a variety of shapes as we will be in *Chapter 19*. Tinkercad's ruler was designed to be moved, reset, and enabled wherever you need to make adjusting dimensions an intuitive task.

When you're done setting dimensions, press the **X** button in the corner of the ruler to remove the tool from your design window. Your measurements will remain, and you can always drag the ruler tool back onto your workplane whenever you need it. Something else you may want to drag into your design is another workplane, which may make things a bit easier.

Using the workplane tool

As we know, a flat workplane is created as the starting point for each 3D design we create. But we also can create workplanes when we want to create a design feature off the typical starting grid. Before using the workplane tool, you will want to start your model so that there is a new surface to design from. For example, I have created a part with an angled surface, as shown in *Figure 6.10*:

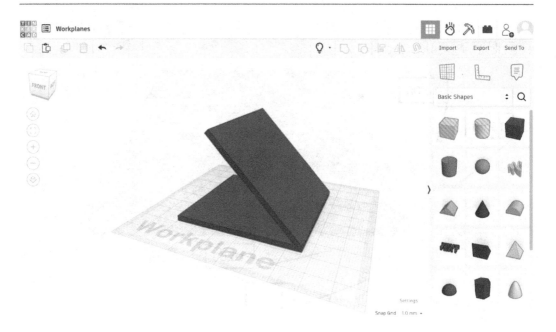

Figure 6.10: An example part with an angled surface

I would again like to put holes on this part so that they are spaced evenly, but I want the holes to be on the angled surface rather than the flat base. Typically, you might drag a hole cylinder onto your workplane, then rotate it and position it to align with the angled surface. Instead, we can use the workplane tool.

After creating a shape, drag the workplane tool from the side panel of the design window onto your new surface. Looking at *Figure 6.11*, we can see how I dragged and dropped the workplane onto the angled surface of my plate:

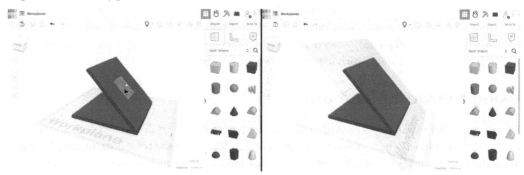

Figure 6.11: Creating a new workplane on a surface

The image on the left of *Figure 6.11* shows the process of dragging a workplane out onto the design, while the image on the right shows what it looks like after the workplane is dropped in place and created. When you create a new workplane, the original blue workplane becomes faded, while the new workplane is shown in orange, as shown in the image to the right of *Figure 6.11*. Now, each new shape that is created will be placed on the new workplane instead of the original one. This allows you to place shapes on unique surfaces more easily, as shown in *Figure 6.12*:

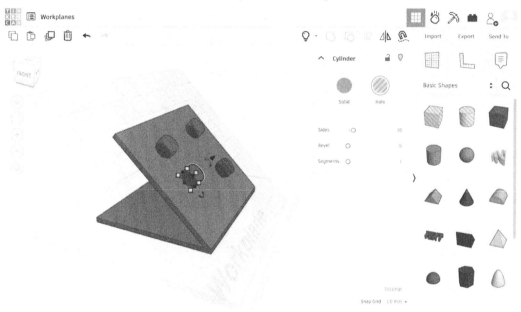

Figure 6.12: Placing shapes onto a created workplane

In *Figure 6.12*, I was able to create cylinder hole shapes and place them on the angled surface by dragging them out onto my created workplane. As the workplane was created on the angled plate, I didn't need to rotate the cylinders to match this angle because this will happen automatically based on the created workplane. The new workplane can also be used to dimension and align our shapes, as shown in *Figure 6.13*:

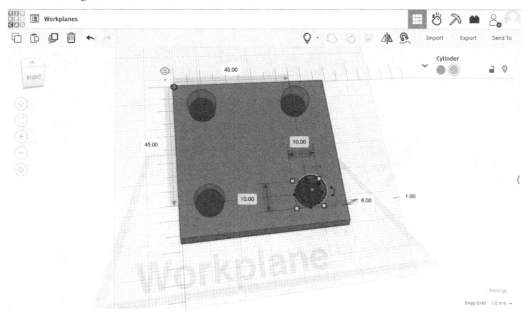

Figure 6.13: Using the ruler tool and workplane tool simultaneously

Like what was shown in *Figure 6.7*, I was able to drag the ruler tool onto my new workplane. With the ruler enabled, I can dimension my hole cylinder shapes accurately in relation to the angled plate they are resting on with the workplane that I created.

When you are done designing on a workplane, or if you want to create a new workplane, you simply need to drag and drop the workplane tool to a new position. Dragging the workplane tool onto the original blue workplane will restore your design window to the default setting, while dragging and dropping it on a surface will create a new workplane to work with. If you want to return to a previously used workplane, like the one that was created in *Figure 6.11*, you will need to repeat the process of dragging and dropping the workplane tool into position.

It's important to note that the workplane tool provides similar capabilities to a new feature in Tinkercad called **cruise mode**. We will cover cruise mode in the following chapter, but a key difference between the workplane tool and cruise mode is that the workplane tool is more suitable for working with multiple shapes on a new surface, as shown in the example within this chapter.

Summary

Looking back at the topics covered in this chapter, it is important to consider how these concepts will be used frequently throughout all the designs you create in Tinkercad. Adjusting the grid size, units, and snap grid increments are settings that can be adjusted each time you create a document, or even as you are working to create a workspace that suits the specific needs of your various projects.

Whenever you are designing parts that will be manufactured or ones that need to fit together, measurements are crucial. Between entering measurements right on the grid or dimensioning parts through the ruler tool, there are multiple strategies that can be implemented to ensure your designs are accurate. And as you create complex shapes, consider using the workplane tool to make designing unique parts not only more accurate, but more efficient as well.

We will continue to reference the tools and strategies introduced in this chapter throughout this book, especially later when we enter the fourth part and begin to design specifically for 3D printing and production. In the meantime, we will continue to dive deep into more tools to create unique and complex designs. When you're ready, flip the page to learn more about the tools that assist with complex multi-part creation.

7
Tools to Manipulate and Pattern Multi-Part Designs

So far, the second part of this book has focused on working with various tools to create and manipulate shapes accurately as we have created more complex 3D designs. In this chapter, we will look at tools that allow us to work with multiple shapes to create complex and aesthetically pleasing designs with greater efficiency.

You may find that the more shapes you add, the more difficult it becomes to navigate through your designs. We discussed this concept earlier in *Chapter 4* when we learned how to hide and lock shapes. In this chapter, we are going to be looking at tools for combining shapes as we create complex designs. We will learn to do so in the context of the following topics:

- Cruising
- Aligning shapes
- Duplicating and patterning shapes

By the end of this chapter, you will not only have learned about new tools that have not been covered previously but also have seen how some previously learned concepts such as dimensioning and grouping shapes can be enhanced through more advanced techniques and some creative thinking.

Technical requirements

We will be using Tinkercad as discussed in the *Technical requirements* sections of *Chapters 1-3*.

An editable model of the patterned pumpkin example shown in this chapter can be accessed on Tinkercad at `https://www.tinkercad.com/things/4JRmjrEkXkm-pumpkin-model-from-chapter-7`.

Cruising

In *Chapter 6*, we learned about the importance of **workplanes**, which are a key starting point for our designs. We also learned how workplanes could be created on surfaces we want to model on as we create complex designs more efficiently.

Now, we are looking at a newer Tinkercad feature called the **Cruise tool**, which has a similar effect to the workplane tool. The first thing we need to do is enable the **Cruise** tool using the icon that looks like a magnet on the top toolbar, as shown in *Figure 7.1*:

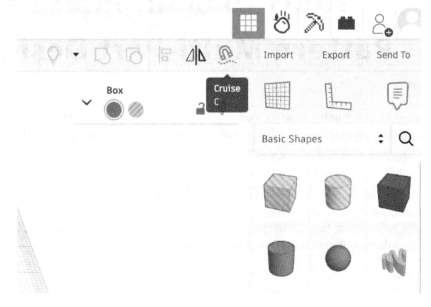

Figure 7.1: The Cruise tool can be enabled or disabled using the icon on the toolbar

We can also turn the **Cruise** tool on or off by pressing C on our keyboard. Once it has been enabled, we will find that we can assemble multi-part designs with greater ease using a snap-like feature. There are two ways to do this. The first is to enable **Cruise** after selecting a shape to move and assemble shapes that are already in our designs, as shown in *Figure 7.2*:

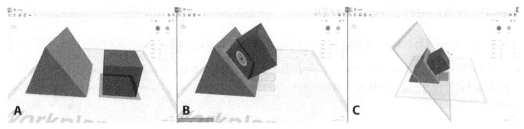

Figure 7.2: Using the Cruise tool to move and connect existing shapes

The image **A** shown in *Figure 7.2* shows what happens after a shape is selected and the **Cruise** tool is enabled by using the icon on the toolbar, or by pressing *C* on your keyboard. We can then use the tool that appears to assemble our shapes through the following steps:

1. After selecting a shape and enabling the **Cruise** tool, click and drag on the dot that appears on the flat surface of your selected shape.
2. You can drag the dot around your design, which will move your shape around as well. You can drag the dot onto a surface, such as the side of another shape, to align your shapes together, as shown in image **B** of *Figure 7.2*.
3. After you release the dot, the shape will remain assembled to this point and a small workplane will be created to show you that the shapes are touching, as shown in image **C** of *Figure 7.2*.

We can also enable **Cruise** as a default option as well. Press the **Settings** button in the bottom-left corner of the screen to open **Workspace settings**, as shown in *Figure 7.3*:

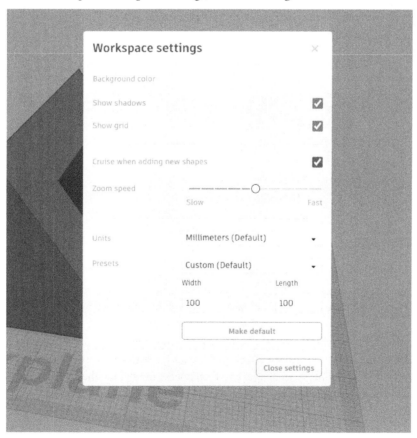

Figure 7.3: Enabling Cruise as a default in Workspace settings

After opening the **Workspace settings** window, check the box next to **Cruise when adding new shapes** to make this the default option. Now when you drag a new shape onto an existing shape with **Cruise** enabled, you will find that the sides of your shapes align and snap to one another, as if there is a workplane between them.

The **Cruise** tool automatically creates a temporary workplane between shapes to make aligning them to one another easy. If you press *W* on your keyboard while you are using the **Cruise** tool, a fixed workplane will be created rather than a temporary one that can disappear. **Cruise** offers a quick way to align and combine shapes, especially when you're trying to do so at odd angles like the ones shown in *Figure 7.2*.

Another keyboard shortcut to use while cruising with shapes is the *Shift* key. The effect of pressing this is demonstrated in *Figure 7.4*:

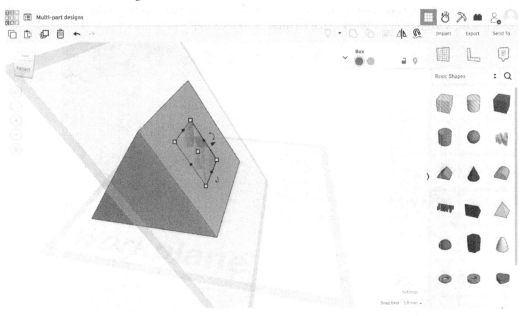

Figure 7.4: Changing the snap surface while using the Cruise tool by pressing Shift

Pressing *Shift* will change which surface is being snapped together when assembling shapes using the **Cruise** tool. By default, the shapes will attach to the exterior surfaces, as shown in *Figure 7.2*. But if you press *Shift* while assembling your shapes and then release the shapes, they will assemble to interior surfaces, as shown in *Figure 7.4*.

It's important to understand that the **Cruise** tool does not group shapes but only allows you to move and align shapes so they are touching. If you would like to make a solid design after aligning shapes with the **Cruise** tool, you would then need to select your shapes and either click the **Group** on the top toolbar, or press *Ctrl + G* on your keyboard.

While the **Cruise** tool is a quick and effective way to connect and create multi-part designs, you may find that the Workplane tool is a more effective way to design if you are combining multiple shapes on the same surface. You may also find that it can sometimes be difficult to align your shapes as you create multi-part designs. Fortunately, there's a tool for that called the Align tool.

Aligning shapes

As discussed previously, there are numerous ways to move shapes around the design space when creating multi-part models. You can simply drag shapes around using your mouse, or you can use the arrow keys to move shapes around as discussed in *Chapter 2*. However, as you may have discovered, aligning shapes perfectly to the left, center, or right edges can sometimes be a tedious task.

Fortunately, there's a tool to assist with this common challenge. It's called **Align**. Prior to using the **Align** tool, you must have at least two shapes in your design and they must both be selected for the **Align** tool to become available, as shown in *Figure 7.5*:

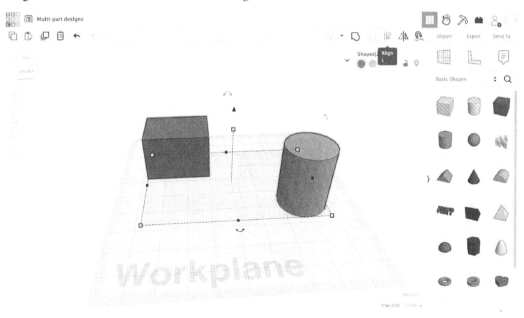

Figure 7.5: Choosing the Align tool to align two selected shapes

With at least two shapes selected, you can enable the **Align** tool using the icon on the top toolbar as shown in *Figure 7.5*, or by pressing *L* on your keyboard. After enabling the **Align** tool, nine black dots will appear on the screen around your selected shapes, as shown in *Figure 7.6*:

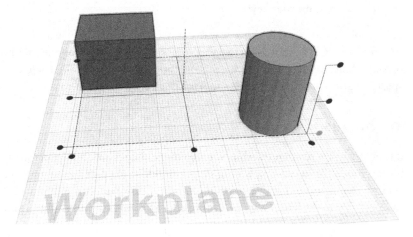

Figure 7.6: Options for aligning shapes using the Align tool

There are three sets of dots, one each for the x, y, and z axis of our design. Within each set, there are three black dots, one for each alignment option. This is similar to how the **Mirror tool** works, which was introduced in *Chapter 5*, though for the **Mirror** tool we could only select which axis we wanted to mirror along rather than also having the option to choose an axis and a position as we can with the **Align** tool.

For example, if I were to click on the middle dot in the set of dots for the horizontal axis (x) across the front of my workplane, my two shapes would become centered to one another, as shown in *Figure 7.7*:

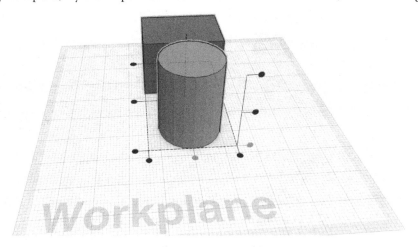

Figure 7.7: The result of aligning two selected shapes using the Align tool

By comparing *Figure 7.6* and *Figure 7.7*, we can see that the shapes moved toward one another in *Figure 7.7*. We can also see that they are centered with one another horizontally, which is now represented by the grayed-out dot shown in the middle set of x-axis dots. They are also aligned to the bottom vertically by default, also shown by a grayed-out dot in the vertical set of dots for the z axis.

So, pressing a *center* dot will center your shapes, and pressing a *side* dot will align your shapes to whichever side you choose. You can choose to align your shapes across multiple axes at the same time, as shown in *Figure 7.8*:

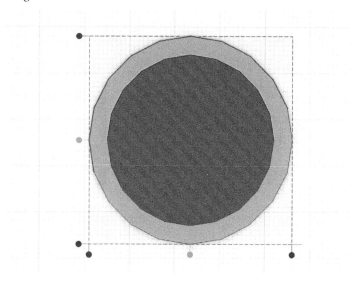

Figure 7.8: Aligning shapes across multiple axes simultaneously

In *Figure 7.8*, I have two cylinders that I want to center on the x and y axis. To do this, I selected both shapes, then selected the center dot for the two flat axes shown in *Figure 7.8*. I didn't want them to be centered vertically, so I ignored the dots on the z axis in the **Align** tool. You will find that the **Align** tool remains on the screen until you click away or deselect your shapes. This allows you to make multiple selections or align your shapes in multiple dimensions with ease.

You may find that using this tool takes some practice and that it can be difficult to determine which black dot or axis you want to align your shapes along. One key strategy to recall is to orbit around the design window so you can view your shapes from different perspectives, as discussed in *Chapter 3*. You can also preview what is going to happen prior to clicking on a dot, as shown in *Figure 7.9*:

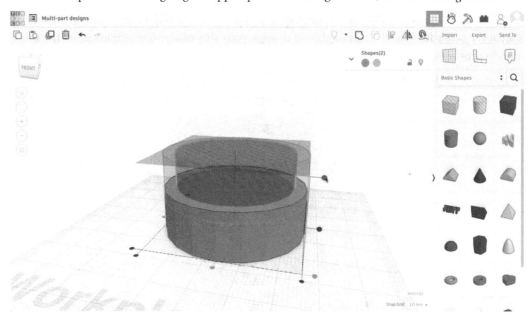

Figure 7.9: Previewing an action using the Align tool

In *Figure 7.7*, my mouse is hovering over one of the dots for the **Align** tool after selecting two shapes and enabling the tool, but I haven't clicked yet. We can see that a transparent shadow for my shapes has appeared to show me what would happen if I pressed on this dot, which allows me to confirm that I am making the right choice before aligning my shapes. But not to worry, you can always press **Undo** if you make a mistake or want to try again!

You may have also noticed that using the **Align** tool moves our shapes across the workplane, as shown in *Figure 7.6* and *Figure 7.7*. In *Figure 7.6*, the two shapes are spread apart, but they move toward one another when aligned in *Figure 7.7*. What if we are aligning our shapes and one of them is in a dimensioned position that we've set and don't want it to move from? Fortunately, there's a strategy for this.

To align shapes to a shape that doesn't move, we can complete the following steps:

1. Select the (two or more) shapes you want to align, then select the **Align** tool.

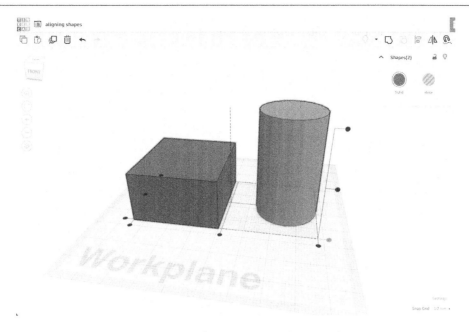

Figure 7.10: Selecting shapes to be aligned

2. With the **Align** tool enabled, click on the reference shape that you want to align to (the shape you don't want to move). As you can see, I clicked on the box shape in *Figure 7.11* to serve as my reference shape:

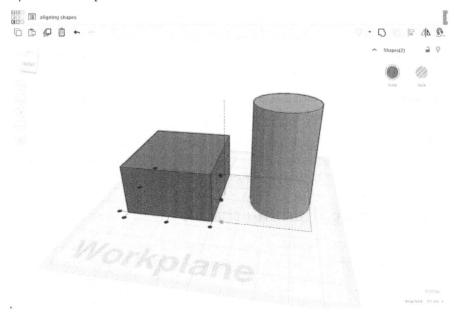

Figure 7.11: Choosing a reference shape to align other shapes to

3. After clicking on a shape, you will notice that the alignment dots are now around the reference shape rather than all the selected shapes. You can then click on the dots for the axis and positions you wish to align to.

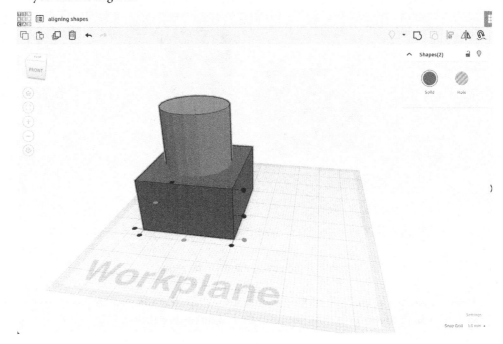

Figure 7.12: Aligning shapes to a reference shape

4. You will see that your other shape(s) move to align to the reference shape based on the dots chosen while the reference shape does not move, as shown in *Figure 7.12*.

We've discussed a lot of different tools and strategies in this book thus far, but the **Align** tool is one to keep in your toolkit and in the back of your mind for every project moving forward.

So far in this book, we've discussed how multiple shapes can be moved, dimensioned, mirrored, attached, and aligned, but we haven't covered one of the simplest actions that can also be one of the most impactful. This is the action of **duplicating** our shapes.

Duplicating and patterning shapes

You might think that discussing copying and pasting shapes isn't necessary to cover in its own section, but duplicating is more than just pressing copy and paste. The **Duplicate and repeat** tool has its own button on the top toolbar next to the copy and paste buttons as shown in *Figure 7.13*:

Duplicating and patterning shapes   101

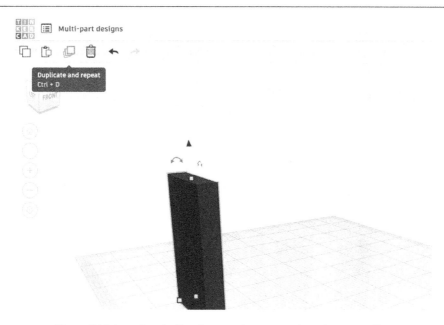

Figure 7.13: Locating the Duplicate and repeat tool on the top toolbar

In addition to using the **Duplicate and repeat** tool via the toolbar button, you can also use the *Ctrl + D* shortcut on your keyboard as well. But before we do, let's establish the difference between using copy and paste and using the **Duplicate and repeat** tool.

If you wanted to make an exact copy of a selected shape or of multiple selected shapes, then you would first press copy, then press paste. Duplicates of each selected shape would then appear on your screen.

Initially, you might find that pressing the **Duplicate and repeat** button does the same thing with one less click (pressing copy and then paste versus just pressing **Duplicate and repeat**). The difference is that the **Duplicate and repeat** tool will not only copy and paste your shapes but also copies and repeats any **shape transformations**.

For example, let's add a rectangular shape and duplicate it so that there are two. You will notice that the second shape is created right on top of the first, as shown in *Figure 7.14*:

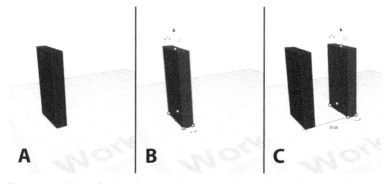

Figure 7.14: Using the Duplicate and repeat tool to make duplicates of a shape

In image **A** of *Figure 7.14*, I have added my rectangular shape to the design. In image **B**, I've pressed the **Duplicate and repeat** tool to make a copy, which is on top of the original. In image **C** of *Figure 7.14*, I've dragged and moved the copy off to the side so that it is 20 mm away from the original.

But what happens if I press **Duplicate and repeat** again? Will a new copy be made on top of the copy created in *Figure 7.14*? No, instead something else happens, as shown in *Figure 7.15*:

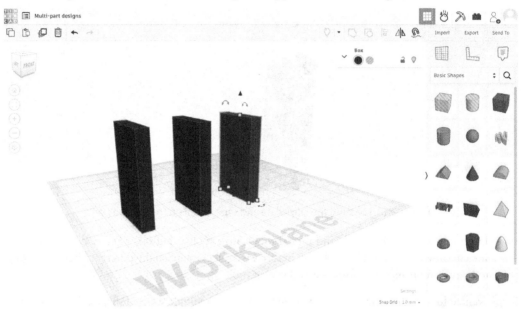

Figure 7.15: Using the Duplicate and repeat tool again on a previously duplicated shape

As shown in *Figure 7.15*, another copy of the rectangle shape was created, but it isn't on top of the shape I duplicated. Instead, it is 20 mm away from it. Why did this happen?

When you duplicate a shape, a copy of both the shape and the transformations made to it will be created. Since I duplicated the rectangle shape *and* moved it 20 mm, the next duplicate was also moved 20 mm. If I were to press **Duplicate and repeat** five more times, I would have a row of rectangle shapes that are each 20 mm apart, as seen in *Figure 7.16*:

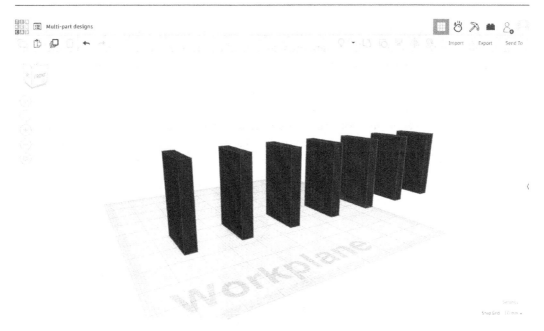

Figure 7.16: Repeating a transformation using the Duplicate and repeat tool

I only moved a single shape 20 mm once, but the **Duplicate and repeat** tool allowed me to repeat this pattern as many times as I wanted to, as shown in *Figure 7.16*. The catch is that I cannot deselect the shape when duplicating, or else the transformation will be lost. For example, if I duplicated the original rectangle shape, moved it 20 mm, then clicked away, then clicked on the moved shape and tried to duplicate it, a new copy would be made on top of it without moving.

There are endless instances where using the **Duplicate and repeat** tool might save you some time in creating a multi-part design, not only by being able to make copies but also by being able to autonomously create patterns in your designs. We can transform our shapes in a lot of different ways, not just through movements. One of my favorites is through rotations, as shown in *Figure 7.17*:

Figure 7.17: Using the Duplicate and repeat tool to create a petal pattern for a flower model

To create the flower petals in the model shown in *Figure 7.17*, I created one long petal centered behind the half-sphere in the middle of the design, then duplicated the petal and rotated it by 45 degrees. By duplicating the petal three more times, I created petals that wrapped around the center part of my flower with perfect spacing and symmetry.

However, the really cool thing about the **Duplicate and repeat** tool is that it will copy all transformations made during your selection, not just a single movement as shown previously. For example, let's see what happens when I move and increase the size of a shape, then duplicate it six times by looking at *Figure 7.18*:

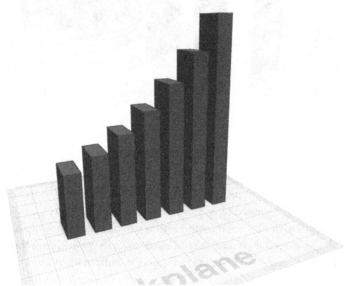

Figure 7.18: Using the Duplicate and repeat tool to make copies of a shape with multiple transformations

Looking at *Figure 7.18*, we can see that the original shape (to the left) gradually increased in size as it was duplicated and moved to the right. This happened because both movement and scale transformation were duplicated with each new shape made.

In the first duplication, I made the second rectangle shape 5 mm taller and moved it 10 mm to the right. The next shape was 5 mm taller and 10 mm farther away from the previous, the one after that was 5 mm taller and 10 mm farther, and so on.

It is difficult to provide examples to show how else the **Duplicate and repeat** tool may be useful, as there are so many open-ended possibilities for it. So, to wrap this section up, let's use the **Duplicate and repeat** tool to create a model of a real-world pumpkin:

1. Start by creating a large oval shape on your workplane by manipulating a **Sphere**, as shown in *Figure 7.19*:

Duplicating and patterning shapes 105

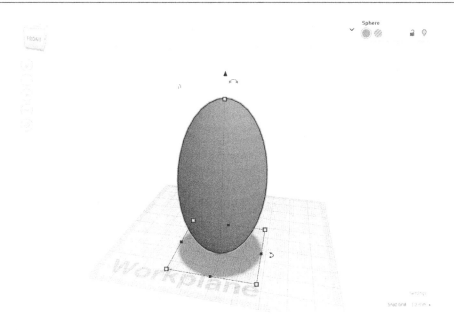

Figure 7.19: Create a sphere shape to start the design of a pumpkin

2. Next, we can use the **Duplicate and repeat** tool to duplicate the **Sphere** shape, then move and rotate it by approximately 25 degrees so it is still touching the first, but off to one side as shown in *Figure 7.20*:

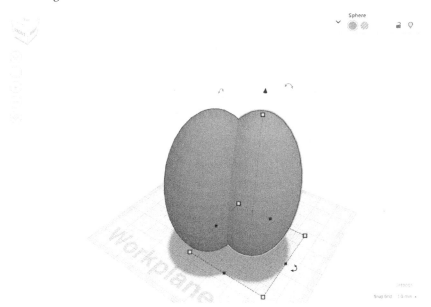

Figure 7.20: Creating the first duplicate and transformations for the pumpkin design

3. Without deselecting the sphere shape we moved, press the **Duplicate and repeat** tool to make another copy. This should move and rotate like the first so it continues off to the side, as shown in *Figure 7.21*:

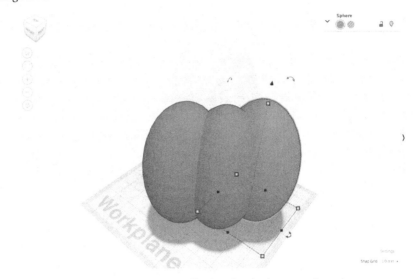

Figure 7.21: Testing and confirming the duplication of our shape

4. Continue to press the **Duplicate and repeat** tool until the sphere shapes wrap all the way around the center sphere, as shown in *Figure 7.22*:

Figure 7.22: Duplicating a pattern to make the body of a pumpkin

5. Add another sphere shape and scale it so it fits in the center of the duplicated sphere shapes, as shown in *Figure 7.23*:

Figure 7.23: Adding a sphere to fill the center of the pumpkin

6. Next, select and drag a polygon shape so that it fits on top of the center sphere shape of the pumpkin, as shown in *Figure 7.24*:

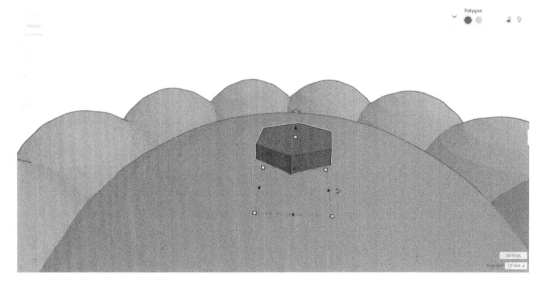

Figure 7.24: Adding a polygon shape to the top of the pumpkin design

7. Duplicate the polygon shape, raise it approximately 3 mm above the original, rotate it to the left by approximately -5 degrees, and make it slightly smaller, as shown in *Figure 7.25*:

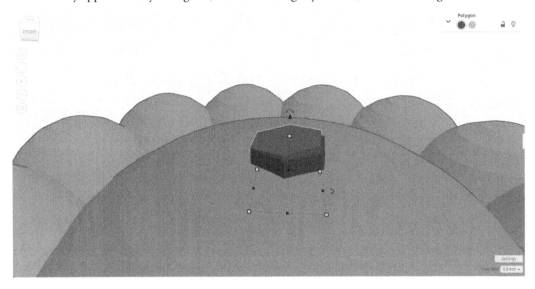

Figure 7.25: Creating the first duplicate and transformations for the polygon shape

8. Without deselecting the new polygon shape, duplicate it repeatedly to create a stem-like shape on top of the pumpkin model, as shown in *Figure 7.26*:

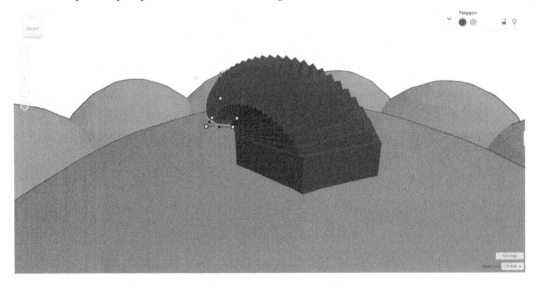

Figure 7.26: Duplicating a pattern to make a stem-like shape

9. Adjust the scale and colors of your shapes as you desire, then group them to create a finished pumpkin model, as shown in *Figure 7.27*:

Figure 7.27: The finished pumpkin design made with the Duplicate and repeat tool

My goal for this mini project was less about teaching you how to make a pumpkin and more about showing you different ways the **Duplicate and repeat** tool can save you time, increase efficiency, and offer opportunities for endless creativity in your modeling.

As you continue to create and expand your wealth of knowledge and understanding for the Tinkercad application, don't forget about the **Duplicate and repeat** tool, as it can cut back on time and allow you to create some amazing shapes along the way.

Summary

Throughout this chapter, we explored new tools and strategies to enhance our designs by manipulating multiple shapes with greater efficiency and accuracy. Using these skills, you may not only find that you can create more unique and complex 3D models but also that you create them more quickly as well.

As we learned in the previous chapter, we can create workplanes on surfaces we would like to model. However, by using the **Cruise** tool, attaching shapes from surface to surface can be done quickly and with only a simple click-and-drag action, as workplanes are created automatically. When you have your shapes in your design and want to position them, you can then use **Align** to accurately align your shapes across three different positions and three different axes.

We also learned that two of the simplest actions, copying and pasting, can be used to create complex patterns with autonomy and ease by using the **Duplicate and repeat** tool. After selecting and transforming a shape, we can duplicate it to not only make a copy but also to repeat the transformation. You can use this to make complex designs such as petals on a flower more quickly, or you could use it to pattern a shape to make a completely unique shape such as the stem of a pumpkin.

Now that we know additional strategies and tools for creating more unique and complex shapes, it's time to learn where we can get and create more shapes! When you're ready, turn the page to learn more about importing shapes and objects into our 3D designs in the next chapter.

8
Importing Models and Designs

In this chapter, we look at answers to the common question, "What if I need more shapes?" As we have seen in earlier chapters, Tinkercad has an extensive collection of shapes that we can manipulate to create unique 3D models and designs, from vases to vises or other real-world prototype solutions. But at times, you may find that there just isn't a shape to meet a specific need or one to enhance your design in the way you are looking to achieve. We will look at how to address these challenges in this chapter through the following topics:

- Tinkering from one design to another
- Importing 3D objects
- Importing vector shapes

Through the topics and skills covered in this chapter, you will learn how to expand what is possible in a 3D design made with Tinkercad by bringing in models, shapes, and designs from other sources to further enhance your creations. By doing this, we will also review and apply some key skills and concepts covered in previous chapters, as well as lead into more advanced concepts that are yet to come.

Technical requirements

Throughout this chapter, we will be looking at bringing models and design files into our Tinkercad designs, which can be obtained from various sources. You can browse public Tinkercad designs in the Tinkercad Gallery at `www.tinkercad.com/things`, which can be copied into your own documents and designs, as discussed later in this chapter.

We can also browse for 3D models in an **STL** or **OBJ** file format, which can be brought into our designs from file-sharing websites such as `www.thingiverse.com` and `www.cults3d.com`. Many of these sites offer models that are free to download and use for non-commercial purposes, and there are others which offer models that you can buy as well. In this chapter, a model of a trophy will be used as an example, which can be downloaded from `www.thingiverse.com/thing:6493928`.

3D models can be created in many different file formats, but Tinkercad is only compatible with STL and OBJ files. You can use sites such as `www.imagetostl.com` to convert various image or design files into an STL file format so that it is compatible with Tinkercad.

Additionally, you can find editable versions of the Tinkercad models shown throughout this chapter at the following links:

- `https://www.tinkercad.com/things/4JRmjrEkXkm-pumpkin-model-from-chapter-7`
- `https://www.tinkercad.com/things/aEUU6ByTOON-imported-trophy-from-chapter-8`
- `https://www.tinkercad.com/things/bsGwMCcwzpI-recycle-basket-model`

We will also look at importing 2D design files in a vector file format, such as a **Scalable Vector Graphic** (**SVG**) image file. SVG images and clipart files can be downloaded for free from sites such as `www.svgrepo.com`, or they can be created with simple and free vector design programs such as Google Drawings (`docs.google.com/drawings`) or `www.vectr.com`. In this chapter, an image file of the recycle symbol will be used as an example, which can be downloaded from `www.svgrepo.com/svg/449408/recycle`.

Tinkering from one design to another

The first place to look for additional design files and models to incorporate into your own project is from within Tinkercad itself. Let's say you are working on a new 3D model and you want to bring in a design that you made previously, such as the pumpkin we created in *Chapter 7*.

To do so, follow these steps:

1. The first thing we need to do is open both (or several) design windows in separate tabs within our web browser, as shown in *Figure 8.1*:

Figure 8.1: Two different Tinkercad design windows are opened simultaneously in tabs

As shown in *Figure 8.1*, I have my pumpkin model open in one tab to the left, and then another design with grass and trees modeled in the tab to the right of my web browser. There are a few different ways to bring the pumpkin model into the garden model, but the easiest is through copying and pasting.

2. From within the pumpkin design, I can select the model by clicking on it or by pressing *Ctrl + A* on my keyboard to select all shapes.

3. After it is selected, I can copy it using the **Copy** button on the top toolbar, or by pressing *Ctrl + C* on my keyboard. I then want to switch to my second design, the grassy model, to paste it, as shown in *Figure 8.2*:

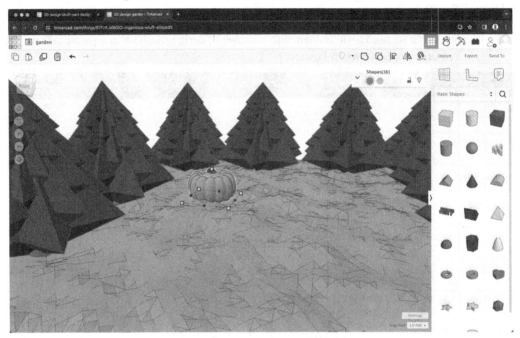

Figure 8.2: Pasting a 3D model from one Tinkercad design to another

4. After switching to my other 3D design, I can bring the pumpkin model that I previously copied in by pressing the **Paste** button on the top toolbar, or by pressing *Ctrl + V* on my keyboard. This will combine the different 3D designs into one, as shown in *Figure 8.2*.

When you copy a design from one Tinkercad model to another like this, all the design's original properties and features remain. This means you can scale, ungroup, and manipulate the shape just like you would in the original design.

But you aren't limited to the models and shapes from within your own design. Tinkercad has a community of more than 80 million users and countless designs that are shared publicly for you to browse and use. You can search through designs by pressing the **Gallery** button on Tinkercad's navigation window, or by visiting www.tinkercad.com/things in your web browser, as shown in *Figure 8.3*:

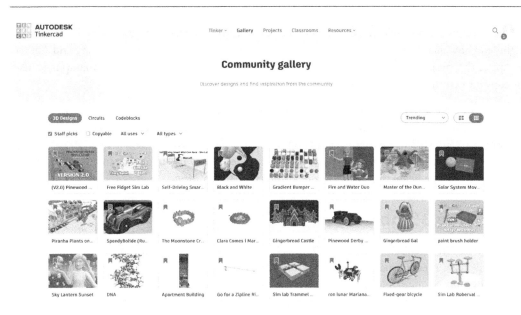

Figure 8.3: Browsing Tinkercad's gallery of public 3D models

Throughout the Gallery, you will find models and designs created by other Tinkercad users, which you can search through by using keywords in the navigation toolbar. As you browse the models, you can select one to view more details or to use in your own designs, as shown in *Figure 8.4*:

Figure 8.4: Viewing the details of a public design from the Tinkercad Gallery

When a design is shared, like the one shown in *Figure 8.4*, you can view the details as well as the usage rights and how many other users have copied, or **remixed**, it previously. By pressing **Copy and Tinker**, a copy of this design will be added to your own Tinkercad designs so that you can manipulate, use, or copy it into other designs, as discussed earlier in this section.

It's important to note that the **Download** button will export the design file in an STL or OBJ file format, and you may also find that some models shared in the Gallery were not made in Tinkercad but instead, imported using these generic 3D model formats. There are benefits and drawbacks to working with downloaded models rather than ones made entirely in Tinkercad, and this is a strategy we can utilize as we begin to consider production using 3D printing.

Importing 3D objects

As mentioned in the previous section, 3D models can be saved in generic file formats such as an **STL** or **OBJ** file type. These formats are universal, and ones that can be shared and opened in many different CAD programs for modeling and production using 3D printers. Tinkercad can export these models, which we will discuss in more detail later in this book as we get into 3D printing our designs, but Tinkercad can also import these models which is a powerful method for enhancing and expanding upon your creations.

You can browse for 3D model files on countless websites, such as *Thingiverse* or *Cults3D*, and many of the models you find are free for non-commercial use. As you browse for different designs, it is important to ensure that it can be downloaded in either an STL or OBJ file format, or else it may need to be converted, as discussed in the *Technical Requirements* section of this chapter. You also need to ensure that the file size is less than 25 MB, or else Tinkercad will not be able to import the model. In this case, you will need to find another program to compress or break up larger files, which can be difficult to do. As such, it is best to import simpler designs into Tinkercad rather than more complex ones.

As an example, I will be downloading and using a 3D model of a pickleball trophy in the following subsections.

Importing a 3D model

To download a 3D model, follow these steps:

1. The 3D model of a pickleball trophy can be obtained from my Thingiverse page, as shown in *Figure 8.5*:

Importing 3D objects   117

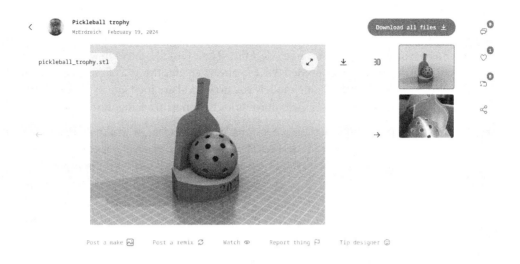

Figure 8.5: Viewing a public design on the Thingiverse file-sharing site

2. After loading this design on Thingiverse, you can find some details about the design, including recommended print settings or comments, and you can also download it in one or multiple parts depending on how it was shared, as shown in *Figure 8.5*.

3. Once downloaded, we can load a Tinkercad design to import the model into a new or existing Tinkercad design, as shown in *Figure 8.6*:

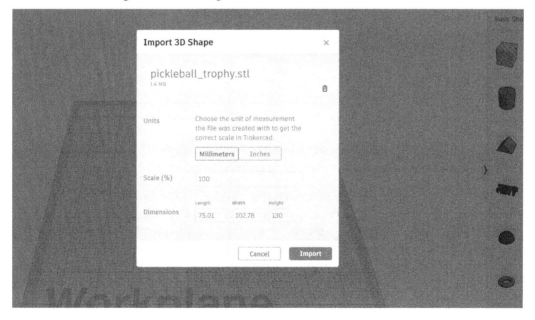

Figure 8.6: Tinkercad's options menu for importing a 3D design file

After pressing the **Import** button in the top-right corner of our design window, we will be shown an options menu to select and adjust our import settings, as shown in *Figure 8.6*. From this menu, you can drag and drop or select the STL or OBJ file you've downloaded previously as well as scale the design as needed. We can always change the size of the models we import later, but you may find that some models you download are too large to fit into your design space. To fix this, you can scale them down prior to importing using this window.

4. Once your imported model has loaded in your 3D design space, which can take a few minutes depending on the complexity of the model being imported, you will notice that it differs from other shapes and models that we have worked with previously, as shown in *Figure 8.7*:

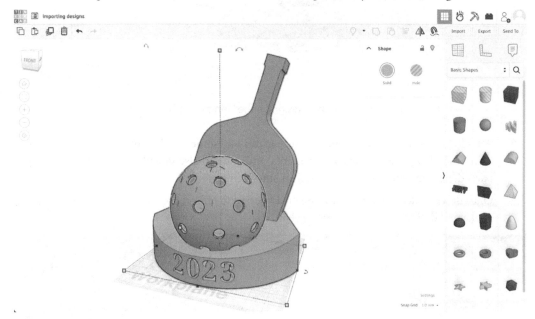

Figure 8.7: Working with an imported 3D model in Tinkercad

5. The color will be automatically set to a random neutral color, but that can be changed using the **Shape** window, as shown in the top-right corner of *Figure 8.7*.

Depending on how the model was created, you may see some cracks or gaps in the imported model, as seen on the ball in *Figure 8.7*, but these small lines aren't typically visible during 3D printing. You will also find that there are no shape parameters to manipulate imported models as with the basic shapes, and that the model cannot be ungrouped. 3D models that are imported in STL or OBJ formats are solid shapes, even if they were separate shapes previously. Once an STL or OBJ file is created, the model becomes a single part that cannot be ungrouped like models created and shared from one Tinkercad design to another.

But this doesn't mean we can't manipulate and use the 3D models we import to create new or unique designs.

Manipulating an imported 3D model

Let's say you have imported this pickleball trophy model, and you want to change the year shown in the model. We can't ungroup this shape to change the text as it was imported, but we can use the fundamentals of CSG to manipulate this design by following these steps:

1. Start by dragging and scaling a **Round Roof** shape onto the front of the trophy model so that the original year is covered, as shown in *Figure 8.8*:

Figure 8.8: Combining basic shapes with imported models in Tinkercad

   As shown in *Figure 8.8*, I used the **Round Roof** shape from the **Basic Shapes** panel (shown to the right of the trophy) to cover the date modeled on the front of the imported trophy design.

2. After adding the round roof shape so that the date is covered completely, I can **Group** the round roof and trophy to make a solid model with no date, as shown in *Figure 8.9*:

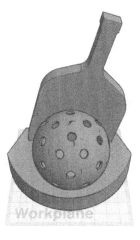

Figure 8.9: Grouping shapes to manipulate an imported model in Tinkercad

As seen in *Figure 8.9*, the date is no longer visible on the trophy because the round roof was grouped and merged with the imported trophy model.

3. We can then use the **Text** shape to create a new date on the trophy, as shown in *Figure 8.10*:

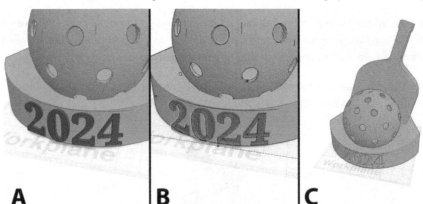

Figure 8.10: Creating number shapes on an imported model in Tinkercad

Image **A** of *Figure 8.10* shows the first step of adding individual number shapes so that they can be rotated to wrap around the front of the trophy model where the original date was covered previously. Next, we can adjust the number shapes so that they are holes rather than solids, as shown in image **B** of *Figure 8.10*. Lastly, we can group the number shapes so that our imported trophy now shows a new date, as shown in the image **C** of *Figure 8.10*.

So, even though an imported model cannot be ungrouped like an original Tinkercad design, all the previously learned topics and skills, such as grouping, creating hole shapes, mirroring, patterns, and adding dimensions can be applied to manipulate and transform imported models too. This allows for imported models to be manipulated so they may become powerful features used to enhance and create our own unique designs.

But importing 3D models isn't all that we can do when importing designs into Tinkercad. I personally find that one of Tinkercad's most powerful features is the ability to import and use 2D design files in a 3D space.

Importing vector shapes

Tinkercad can import and export vector image files in an **SVG** format like how STL and OBJ files can be used for 3D models, as discussed in the *Importing a 3D model* section earlier. **Vector** images are images that are created using points and paths, which differ from the more commonly used **raster** image files made from pixels, such as photographs. Vector image files are widely used in the industry for logos, artwork, and production, and we can take advantage of these files to enhance our 3D designs in Tinkercad as well.

But before you look to incorporate a vector image into a Tinkercad design, it is important to identify where vector images can be obtained from, as well as ensure that they are vector files that will work well with Tinkercad.

Creating and importing custom vector image files

You can create your own vector images using a wide range of programs; I find the simplest are *Google Drawings* and *Vectr*. One key reason to make your own vector file is to create text using fonts that may not be available in Tinkercad through the basic shapes or shape generators, as discussed in *Chapter 5*.

As an example, I have created a textbox using a unique font in the Google Drawings program, as shown in *Figure 8.11*:

Figure 8.11: Creating text to export as a vector file in Google Drawings

Once my textbox has been created in Google Drawings, I can export this design as an SVG file. Like STL and OBJ files, an SVG cannot be edited in Tinkercad, just manipulated. This means I will not be able to change or edit the text once the SVG file has been exported and brought into Tinkercad.

After exporting, I can open a new or existing Tinkercad 3D design file to which I want to import this SVG file. I can then press **Import** to open the import options menu, which we have seen previously, as shown in *Figure 8.12*:

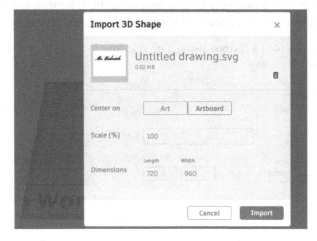

Figure 8.12: The import options menu for importing a vector image into Tinkercad

As shown in *Figure 8.12*, we will find that we can adjust the scale or size of the design in the options menu once an SVG file has been selected, though this can always be adjusted later. After importing the image file into our 3D design, we will see that a random color has been chosen and height was added to the design so that it becomes a 3D model, as shown in *Figure 8.13*:

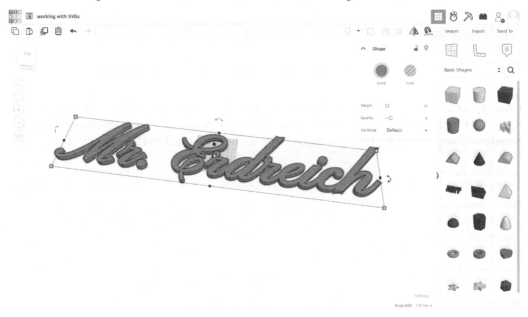

Figure 8.13: An imported SVG image in Tinkercad

SVG files are flat 2D images. But when imported, Tinkercad adds height so that they become flat 3D models instead, as seen in *Figure 8.13*. Like any shape, we can manipulate the size, rotate, mirror, and adjust the color, but we cannot edit the text or the original file once it has been exported and imported as an SVG.

You will also notice that **Fill Mode** can be changed using the shape parameters shown in the top-right corner of *Figure 8.13*. SVG images can be displayed in a few different ways, as shown in *Figure 8.14*:

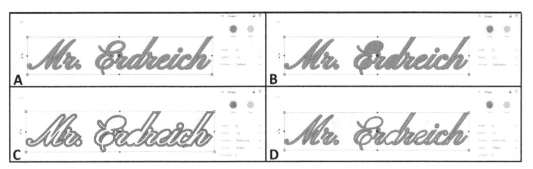

Figure 8.14: Adjusting Fill Mode for imported vector image files

Image **A** of *Figure 8.14* shows the **Default** display option, which fills the shape, while image **B** shows the **Silhouette** display option, which fills the shape, including any holes within the body of the shape. Image **C** shows the **Outer Line** option, while image **D** shows what happens if we select **Inner Line**. All four options use the same image file, but each one offers a different effect and appearance. You can use this parameter to adjust the imported image so that the model appears as you intend it to, and you can even adjust the corners to **Sharp**, **Round**, or **Flat**. This is only possible because an SVG file is made from points and paths, and Tinkercad can manipulate basic parts of these paths through the shape parameters.

Browsing the web for vector image files

In addition to creating and importing our own custom vector image files, we can also browse the web for image files that are free for non-commercial use. One of my favorite sites to use is *SVG Repo*, which includes an extensive library of clipart and image files compatible with Tinkercad. As you browse for shapes through SVG Repo's library, you want to look for **Monocolor** shapes, as shown in *Figure 8.15*:

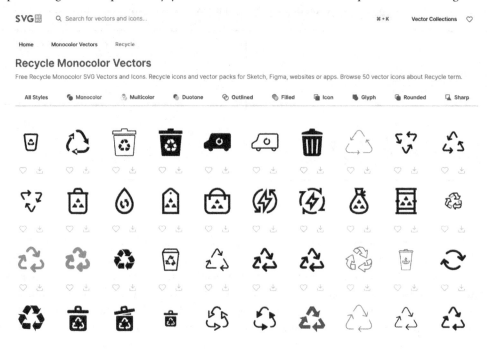

Figure 8.15: Browsing SVG Repo for Monocolor vector image files

As seen in *Figure 8.15*, vector image files can be created and downloaded in single or multi-color designs, as well as varying levels of complexity. As you browse through image files to import into your Tinkercad designs, there are a few key strategies to consider for higher levels of success:

- **Look for single-color images** – The images you choose do not need to only be black but can be any single color. As the images are imported into Tinkercad, all colors will be merged into one so a multi-color design may lose much of its original detail.

- **Look for simple images** – You may find complex vector designs and images that look more like photographs than clipart, and you may want to include these in your 3D models. If you are designing models that will remain on the screen, this is a great way to enhance your designs. But if you plan to import an image to create a model that may be manufactured using a 3D printer, complex artwork and fine details may not be able to be manufactured effectively. Stick to more rounded corners rather than sharp angles when possible.

- **Combine images** – As you look for the perfect artwork to enhance your design, you may find that combining multiple designs in Tinkercad is more effective and easier than trying to find a single image to work with. By importing multiple SVG files, you can easily scale, group, manipulate, and adjust the color for each image using individual shapes in Tinkercad. This is something we'll look more at later on in *Chapter 18*.

As an example, I've downloaded an SVG image of the symbol for recycling using the link provided in the *Technical requirements* section of this chapter. Once downloaded, I imported the SVG image into a Tinkercad design, as shown in *Figure 8.16*:

Figure 8.16: Importing an SVG image from SVG Repo into Tinkercad

As we can see in *Figure 8.16*, the imported image is assigned a random color, and it is also quite large compared to the rest of my design. But I can scale the image as well as rotate it to position it on the model of a wastebasket. By using some of the key tools we discussed in early chapters, such as **Cruise** and **Align**, I was able to reposition and adjust the color of the imported image to create my final model, as shown in *Figure 8.17*:

Figure 8.17: A simple 3D model enhanced with the addition of an SVG image

As shown in *Figure 8.17*, importing a simple image to add to our 3D designs can really work to create realism and detail for our 3D models, which would be difficult to achieve using only the tools provided in Tinkercad. What was a simple rectangular box is now very clearly a recycling bin that I can use in a 3D space I am creating, or even export to 3D print for my desk.

As we have learned how to copy and import models and shapes, you may now be wondering how you could create your own shapes. There is certainly more to learn here and this is something we will look at more closely in the following chapter.

Summary

While this chapter was a bit shorter than some of the other ones we've looked at in this part of the book, it is one that opens a range of possibilities far bigger than what's possible in Tinkercad alone. As you create your own unique designs, or as you browse the designs created and shared by other Tinkercad community members, we now know how we can bring designs from one design space to another by simply copying and pasting.

We've also learned that models exported from Tinkercad or other CAD programs can be imported as STL or OBJ file types, and there are also ways to convert files into these universal formats using public tools found on the web. While we can't edit an imported model like we can with a model created within Tinkercad, we can still apply previously learned skills to manipulate and transform imported models to enhance our own unique creations.

These imported designs don't need to be 3D; they can also be 2D images in the SVG file format. Using imported SVG files, we can add further details and features to our designs that couldn't easily be created using the existing tools that we've discussed in previous chapters. And now that we've gained proficiency in modeling in Tinkercad and have learned how to create custom models using imported designs, we're going to look at how to turn these models into custom shapes in the next chapter. When you're ready, turn the page to start creating your own custom shapes!

9
Making Our Own Shapes

Before moving on to the next part of this book, there are a few things we still need to cover when learning advanced strategies for working with Tinkercad. In this chapter, we are going to combine some of the skills and concepts covered in previous chapters as we strive to create our own custom shapes.

As you continue to design unique 3D models in Tinkercad, you may find that you use similar strategies and methods frequently. You may also find that after importing models or designs into Tinkercad once, as discussed in *Chapter 8*, there is a need to reuse shapes from one design to the next.

We will be looking at how these things can be achieved, as well as what other resources exist to make our abilities in designing complex 3D models that much more effective through the following topics:

- Scribbling shapes
- Working with the Shape Generators
- Creating custom shapes

By the end of this chapter, you will not only be aware of some new tools and strategies that can enhance your creations but also how to be a more proficient creator and 3D modeler in Tinkercad as a whole!

Technical requirements

While there are no new technical requirements for this chapter, as we will be using the same technical requirements listed in *Chapters 1-3*, you can find an editable version of the example model shown throughout this chapter at this link: `https://www.tinkercad.com/things/1QjClefgEoF-creating-shapes-in-chapter-9`.

Scribbling shapes

Earlier in *Chapter 4*, we discussed the differences between Tinkercad and a more traditional **parametric** approach to modeling in CAD software. During this, we discussed how in a parametric CAD program, you typically create 2D sketches, which are turned into 3D shapes rather than starting with a 3D shape as we do in Tinkercad.

In *Chapter 8*, we looked at how 2D **SVG** images could be created and imported into Tinkercad, which would allow you to utilize sketching tools in another program to draw unique designs that could not be done using Tinkercad's tools and features.

Tinkercad may one day add a parametric sketching feature to its 3D design editor, but for now, we can take advantage of a shape called **Scribble**.

Scribble shape basics

As briefly shown in *Chapter 4*, the **Scribble** shape allows you to free-draw shapes in Tinkercad's 3D editor. While this may not be as refined as a vector drawing feature you can find in other CAD programs, it does still have plenty of uses when striving to create your own unique shapes.

The first step to using the **Scribble** shape is to drag and drop the **Scribble** shape onto your **Workplane**, as shown in *Figure 9.1*:

Figure 9.1: Dragging the Scribble shape onto the workplane

As shown in *Figure 9.1*, we can select and drag the **Scribble** shape onto the **Workplane** as we can with any other shape. But once we drop the **Scribble** shape, a new window appears, as shown in *Figure 9.2*:

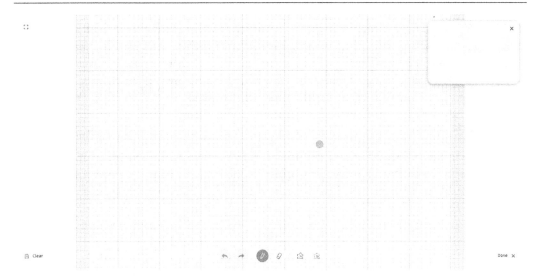

Figure 9.2: The drawing window for the Scribble shape

In this new window, we are presented with some simple drawing tools that can be used to draw a custom shape. Let's take a look at them one by one.

Draw tool

The default tool is **Draw**, which allows you to draw a shape using a tool that is like a paintbrush, as shown in *Figure 9.3*:

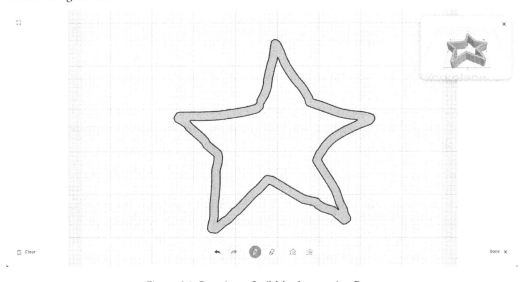

Figure 9.3: Drawing a Scribble shape using Draw

This method is great for drawing small things, or for filling in details, but it's difficult to use when attempting to draw a large shape such as the one shown in *Figure 9.3*. After releasing your mouse to complete the action of drawing a shape, you will see a preview of what that shape looks like in 3D in the top-right corner of the screen, as shown in *Figure 9.3*. As with importing SVG images, as done in *Chapter 8*, the drawn scribbles will be flat 3D shapes when created.

Draw Shape tool

You may find that it is difficult to draw a larger shape with the **Draw** tool. Therefore, as an alternative, I prefer to use the **Draw Shape** tool, which works a bit differently, as shown in *Figure 9.4*:

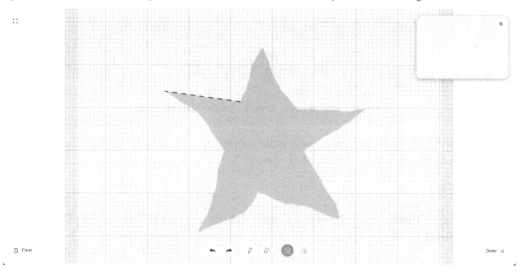

Figure 9.4: Drawing a Scribble using Draw Shape

As seen in *Figure 9.4*, the **Draw Shape** tool allows you to draw a filled-in shape based on the perimeter created by clicking and dragging your mouse across the window. This takes some getting used to, but it is a quick way to create a shape that is entirely filled in, such as the star shape shown in *Figure 9.4*. After drawing a shape with **Draw Shape**, you may find going back to adjust what you drew with the **Draw** tool is an effective approach.

Editing tools

Two different eraser tools mimic the functionality of the **Draw** and **Draw Shape** tools, but remove shapes rather than create them. You will also see a button labeled **Clear**, which clears the entire design if you want to start over, and a button to finish scribbling and return to the 3D editor labeled **Done**. Once pressed, we will see the shape we drew on the **Workplane**, as shown in *Figure 9.5*:

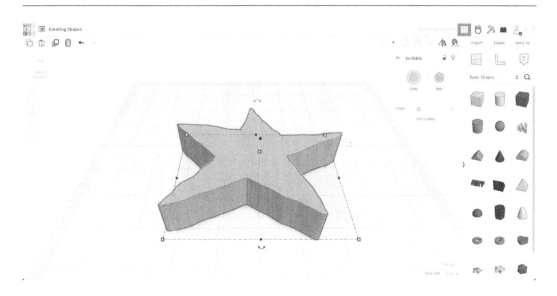

Figure 9.5: A completed Scribble shape

Like any other basic shape, the **Scribble** shape we create can be moved, scaled, rotated, and grouped in our design. As seen in *Figure 9.5*, the shape parameters differ for scribbles though, only with options to adjust the **Height** value or edit the Scribble, which will take you back to the drawing window to adjust the shape's design.

If you are looking to create perfectly smooth lines and curves, the **Scribble** shape is probably not the best approach. For that, combining shapes through the principles of CSG or importing an SVG file might still be the better option. Or, using a **Shape Generator**, but we'll get more into that later in the *Working with the Shape Generators* section.

Where I find the **Scribble** shape to be most useful is when combined with other shapes to create a more organic design, as described in the next section.

Putting the Scribble to use

As the **Scribble** tool is really a free-form approach to designing shapes in Tinkercad, it truly is impossible to define what can be created with this tool. However, as there are so many other shapes to use, it may also be difficult to see where the **Scribble** tool might fit in.

Making Our Own Shapes

To support this, let's complete a simple project using some of the other things we've learned about thus far in this book.

1. First, let's use the **Scribble** tool to draw a curvy organic shape, like the one shown in *Figure 9.6*:

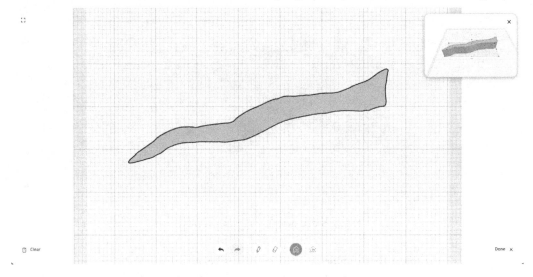

Figure 9.6: Drawing an organic curve using the Scribble tool

2. After pressing **Done**, we can increase the height of the drawn shape and rotate it 90 degrees, as shown in *Figure 9.7*:

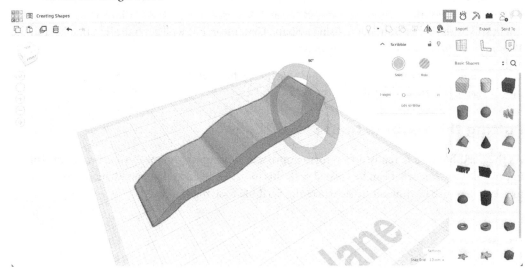

Figure 9.7: Changing the size and position of the drawn shape

3. Next, let's drag a **Roof** shape into the design that is set to be a **Hole**, and position the **Roof** shape along an edge of the drawn shape we created in *Step 1*, as shown in *Figure 9.8*:

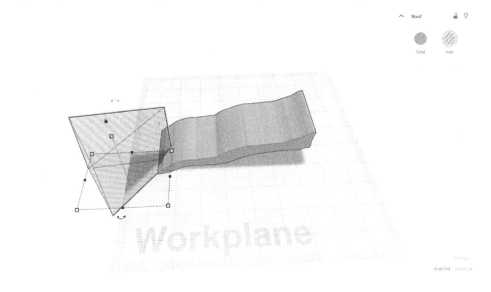

Figure 9.8: Adding a Roof shape to our design

4. As we are striving to create something a bit more organic, you can copy and paste the **Roof** shape somewhat at random, then place the copies around three sides of the drawn shape, as shown in the screenshot on the left of *Figure 9.9*.

5. Once placed, you can then group all the **Roof** shapes with the **Scribble** shape to create the result shown on the right in *Figure 9.9*:

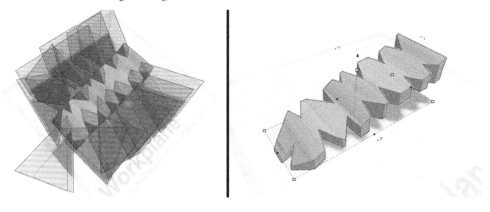

Figure 9.9: Placing Hole shapes around the edge of the Scribble, then grouping the shapes

6. At first, the grouped shape may not look like much, or anything really. But this custom shape we just created has a lot of potential! What does it look like if we set the color to green, and then use the **Duplicate and repeat** tool (as discussed in *Chapter 7*) to make four copies that revolve in a circle around the original shape? You should be able to create something like what is shown in *Figure 9.10*:

Figure 9.10: Making a pattern using the grouped shape

7. These patterned shapes can then be grouped and scaled down to create the top of a palm tree design! We can then move this out of the way for now and add a **Paraboloid** shape to our design to start the trunk of a tree, as shown in *Figure 9.11*:

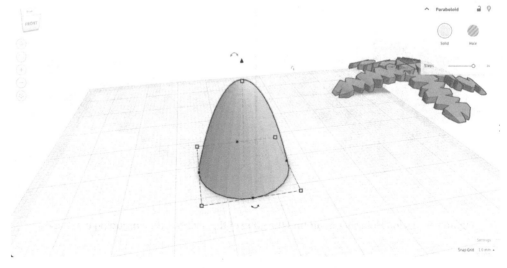

Figure 9.11: Adding a Paraboloid shape to the design

8. Next, use the **Duplicate and repeat** tool to make a copy of the **Paraboloid** shape. You can then raise the copy up, scale it down slightly, and rotate it by 3 degrees. The exact dimensions for this transformation aren't crucial; just ensure the copied shape remains selected as shown in *Figure 9.12*:

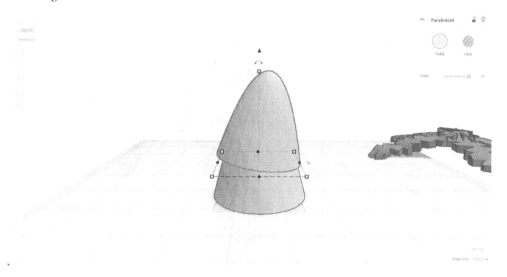

Figure 9.12: Transforming a copy of the Paraboloid shape

9. We can then use the **Duplicate and repeat** tool to make and transform six more copies of the **Paraboloid** shape, as shown in *Figure 9.13*:

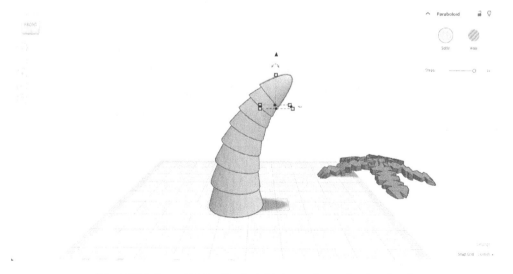

Figure 9.13: Grouping the Paraboloid shapes to create a tree trunk

10. Lastly, we can group the **Paraboloid** shapes and set the color to brown to finish a tree trunk-like shape.

11. We can then bring the grouped **Scribble** shapes back into our design and connect them to the top of the grouped **Paraboloid** shapes to complete a palm tree design.

12. You can also bring in some **Sphere** shapes and add them to the design if you want to create something that looks like coconuts too, as shown in *Figure 9.14*:

Figure 9.14: Completing the palm tree design

While this little palm tree project might seem random, I often find that the use of the **Scribble** tool can sometimes be a little random too. As such, it's a handy tool to keep in mind whenever you find yourself trying to figure out how to make a design turn out the way you want it to and the other shapes available just aren't quite cutting it.

Another handy collection of shapes to support this is called **Shape Generators**, which we are looking at in the next section.

Working with Shape Generators

As initially introduced in *Chapter 5*, **Shape Generators** in Tinkercad is a collection of customizable shapes that can be found in the **Shapes Library**, as shown in *Figure 9.15*:

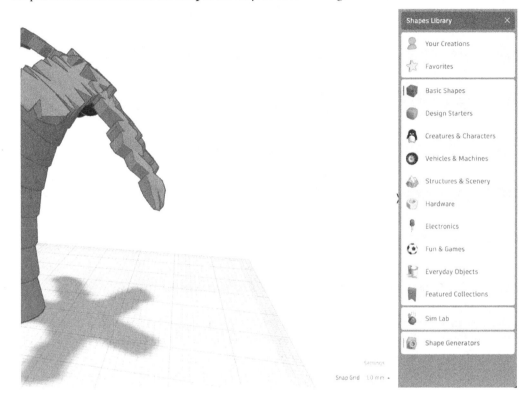

Figure 9.15: Selecting the Shape Generators in the Shapes panel

The shapes in the **Shape Generators** section are unique from the other shapes you can find in Tinkercad for a few different reasons. First, you will notice that each shape generator has vastly different parameters to adjust the shape from one to the next. This allows for a high level of customization as you strive to make unique projects.

You will also note that the shapes have authors, as at one point Tinkercad community members were able to design and contribute shapes through **Shape Generators**. While this is no longer an option, you can use **Codeblocks** to design custom shapes using programming techniques, as mentioned in *Chapter 1*.

The collection of shapes in the **Shape Generators** library is also very random, unlike the other categories we can browse from. You'll find text shapes, like the ones we looked at in *Chapter 5*, and shapes to make gears or sockets, curves, springs, QR codes, graphs, or even a stone wall!

Because of this, I recommend you take some time to browse and play with the **Shape Generators** library so that you become a bit more familiar with what's available. This might allow you to find some useful sources of inspiration for a project you've been working on, such as adding a curvy surface to represent sand in the palm tree design we made earlier:

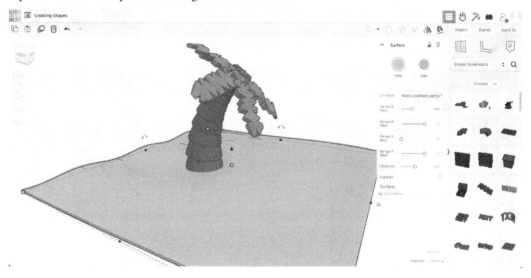

Figure 9.16: Incorporating the Surface Shape Generator into our design

The **Surface** shape, shown in *Figure 9.16*, allows you to enter an equation to create a unique surface, which is an interesting way to model your design. There are several **Shape Generators** that work like this, and others work more like the **Scribble** tool, as shown in *Figure 9.17*:

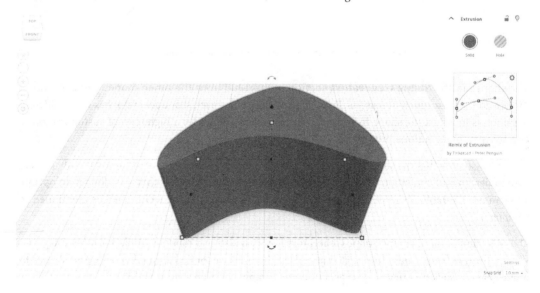

Figure 9.17: Using the Extrusion Shape Generator

The **Extrusion** shape, shown in *Figure 9.17*, starts as a simple cylinder when added to your design, but has a drawing window in the shape parameters. This lets you design a unique *Bezier* curve for your shape. This offers a similar feel to working in some traditional graphic design programs and really allows for a unique and more precise approach to creating custom shapes in Tinkercad.

When striving to create technical projects suitable for 3D printing, I also find the various gear generators to be particularly handy. Another one of my favorite generators is the **Snap and Socket** shape, shown in *Figure 9.18*:

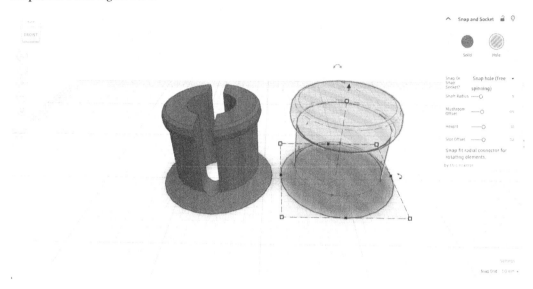

Figure 9.18: Using the Snap and Socket Shape Generator

As shown in *Figure 9.18*, the **Snap and Socket** shape really allows you to make two shapes, a snap and a socket. This is handy if you want to 3D print a multi-part design that connects, like the joint of a toy action figure, spinning wheels, or even the latch of a box. What I like about this shape is that it considers **tolerances** for 3D printing, which is something we talk more about later in *Chapter 12*. This means that you can design with the **Snap and Socket** shape so that the fit between the parts is tight or loose and free-moving, all with some simple adjustments in the shape parameters.

Like the **Scribble** tool, it's difficult to say when exactly the **Shape Generators** may be needed to complete a project. However, there are a wide range of handy shapes, which can help create unique designs in countless circumstances. Also, when you're finished making a unique design, you might find yourself wanting to create a custom shape so that it can easily be reused later, which is something we are going to learn how to do in the next section.

Creating custom shapes

Custom shapes differ from creating just another design in Tinkercad as the shapes you create will be accessible in the shapes library, just like any other shape we've used thus far. You can also turn any 3D design into a custom shape, whether it be made from a single shape or a collection of shapes that have been grouped together.

As an example, I've isolated one of the leaves from the palm tree design we made earlier and selected **Your Creations** in the shapes library, as shown in *Figure 9.19*:

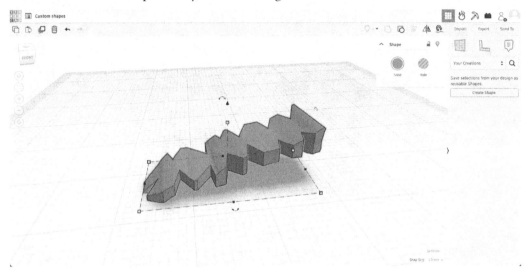

Figure 9.19: Opening Your Creations in the shapes library

Your Creations is where custom shapes can be created and accessed in all of your Tinkercad designs. To create a custom shape, you can select the part or parts of your design you wish to turn into a custom shape, then click **Create Shape**. This will bring up a new window, as shown in *Figure 9.20*:

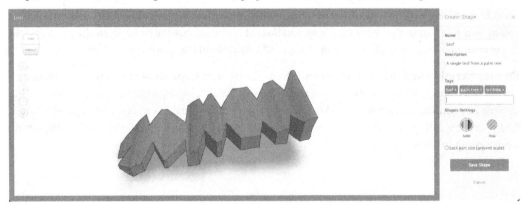

Figure 9.20: Options for creating a custom shape

In the **Create Shape** window shown in *Figure 9.20*, we have a few options we can complete when creating a custom shape. We can name the shape, as well as enter a description or tags, which will make it easier to search for. We can also choose whether this shape should be a **Solid** or **Hole** by default, as well as whether it will be scalable. If you choose **Lock part size**, you will not be able to change the size of it in any future design. This is handy if you design something that is of a specific size, such as the mobile device we will be modeling later in *Chapter 20*. But for something like the leaf I am making, this most likely wouldn't be a suitable option to choose.

After creating your custom shape, you can drag and drop it into your design from the **Your Creations** collection of shapes, as shown in *Figure 9.21*:

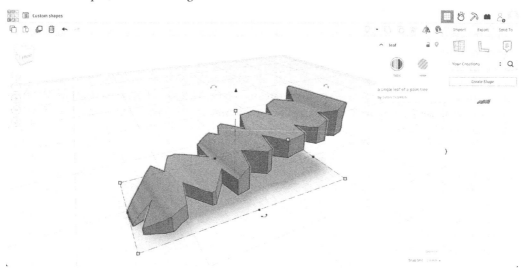

Figure 9.21: Using a custom shape in our design

As seen in *Figure 9.21*, the parameters for this custom shape differ a bit from what we've seen in the past, with the description of the shape listed and yourself labeled as the creator. You will also notice that this custom shape cannot be ungrouped, even though I know the original leaf, shown in *Figure 9.19*, consisted of a scribbled shape and some roof shapes grouped together. Whenever a custom shape is made, it becomes a new solid part, and any shapes used to make it will no longer be editable. The only way to edit the makeup of a custom shape is to keep a copy of the original shape used in a Tinkercad design.

Looking back, we've learned many different ways to create unique shapes in our Tinkercad designs. In previous chapters, we learned how to import models or SVG images, and now we know how to use the **Scribble** shape and various shapes from the **Shape Generators** library, too. As you strive to make unique things moving forward, remember that there is also a way to save and reuse them in **Your Creations**.

Summary

While this chapter touched upon some concepts previously discussed in the earlier chapters, it has also laid a foundation for many more advanced topics that are yet to come.

As you strive to be a designer and one who works to manufacture uniquely designed creations using Tinkercad and 3D printing, it's important to know about all the tools you have at your disposal. I truly believe anything can be created in Tinkercad, and that's largely because of the different tools we can use to make unique shapes.

When you find yourself trying to combine shapes or make a project that reflects something you've seen in the real world, remember that there are tools to draw shapes, and also generators to make unique things based on various parameters. These tools, in combination with the ability to save and reuse your unique creations, will allow you to become a more proficient and advanced Tinkercad user in the chapters to come.

We will also put many of these skills to work later as we utilize **Shape Generators** again in future projects, as well as learn how to manufacture our designs in the next part of this book. Turn the page when you're ready to get started with 3D printing our Tinkercad creations!

Part 3: Designing 3D Models for 3D Printing

As we enter the third part of this book, we will now begin to focus on how our Tinkercad designs can be brought off the screen. Starting with an introduction and overview of what 3D printing is, how it works, and comparing commonly used 3D printing techniques, we will then look at how designs in Tinkercad can be manufactured with 3D printers. As we have learned thus far, nearly anything can be designed in Tinkercad. However, we will quickly learn that different considerations and strategies must be employed if we want to be able to 3D print our Tinkercad designs so they look and function as intended.

This part includes the following chapters:

- *Chapter 10, An Introduction to 3D Printing and Production Techniques*
- *Chapter 11, General Strategies for Creating Effective Models for 3D Printing*
- *Chapter 12, Creating Tolerances for Multi-Part Designs*
- *Chapter 13, Design Mistakes to Avoid*
- *Chapter 14, Exporting and Sharing Tinkercad Designs for Manufacturing*

10
An Introduction to 3D Printing and Production Techniques

With this chapter, we start the third part of this book, which focuses on bringing our designs to life through 3D printing. Over the next five chapters, we will look at strategies and techniques specific to 3D printing while utilizing what we learned previously, including skills and strategies to design with Tinkercad.

Starting with this chapter, we will cover the basics of 3D printing as we look at how it works and how it is used in industry, as well as compare different options and techniques available for our own design and fabrication uses. To do this, we'll look at the following topics:

- What is 3D printing?
- How does 3D printing work?
- Comparing 3D printing techniques
- Choosing the right material

These topics will allow us to cover important terms and concepts before progressing to more hands-on applications and strategies in the following chapters. By the end of this chapter, you will be familiar with the key strategies used for 3D printing, as well as have a greater understanding of how this revolutionary production technique can be used to bring your own 3D models to life!

Technical requirements

Throughout this chapter, the *3DBenchy* 3D model is referenced and used to compare the quality and performance of different 3D printers and 3D printing production techniques. You can obtain this model and learn more about it at www.3dbenchy.com.

As this is an introductory chapter, intended to provide key background concepts and topics before we move on to more practical applications later, there are no technical requirements needed to ensure success. However, access to a 3D printer will be needed to manufacture your designs using the concepts covered in this chapter.

You do not necessarily need to have a 3D printer of your own however. In *Chapter 14*, we will look at 3D printing services available to the public, as well as how to send our Tinkercad designs to 3D printers using the concepts introduced in this chapter. As you prepare to do this, it is important to identify the type of 3D printing production technique you want to use, as well as the material best suited for your design. These topics will be introduced and outlined in this chapter and continue to be referenced later on in the book.

But first, let's begin this chapter by answering a very important question – what is 3D printing?

What is 3D printing?

If you're reading this book, there's a chance that you already have been introduced to 3D printing in some way, but perhaps not necessarily one that covers the full scope of this amazing technology, how it works, and the different ways that it is used.

The first thing we need to know is that there are many different types of 3D printers, and I don't just mean different brands. As we will learn, some 3D printers use liquid, others melt material, and some even construct models using living cell tissue! We'll dive deeper into some of these concepts later, but it's important to note that this is an incredibly diverse production technique that expands every year.

Regardless of which type of 3D printer you find yourself using, all of them essentially perform the same task of taking a digital design file and turning it into a physical model through **additive manufacturing**. Additive manufacturing is the process of creating a product by adding material, and this differs from other manufacturing techniques, as shown in *Figure 10.1*:

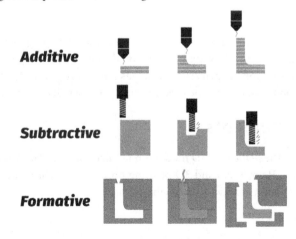

Figure 10.1: Comparing different manufacturing techniques

All 3D printers create models by adding material, usually some form of plastic, and this is typically done by adding material one layer at a time. While many 3D printers use plastic as the production material, there are ones that can create models using clay, metal, chocolate, pancake batter, and even cell tissue! We'll dive deeper into some of these material choices later.

Figure 10.1 shows that additive manufacturing is just one approach to creating a product. **Subtractive manufacturing** is the inverse approach, one where you would insert a block of material, such as wood or metal, and the machine would then carve away excess material to create your product. This is typically done using a CNC (computer numerical control) milling machine, although there are many different options. **Formative manufacturing** is another technique and also the most widely used one to manufacture the products we buy and use every day. As shown in *Figure 10.1*, a product made through a formative technique is done by pouring some type of material into a mold, and the mold then shapes, or forms, the final product.

Each of these production techniques has its benefits and drawbacks, as well as uses in both prototyping and industrial applications. 3D printing has become an increasingly popular choice, thanks to its versatility, ease of use, and lower cost for lower volume production. Every year, 3D printing evolves a bit more, with the release of new 3D printers that can work with new materials and in even greater detail or speed. Setting up and using a 3D printer is also typically easier to do in additive manufacturing when compared to a CNC mill for subtractive manufacturing. And while mass-producing a part with molds through formative manufacturing may be quick and cost-effective, 3D printing is far more affordable for small-quantity production or when creating individually unique parts.

While 3D printing has grown exponentially over the past two decades, it is still predominantly used for **prototyping**, and a 3D printer is often categorized as a **rapid-prototyping** machine. Before a company invests tens of thousands of dollars in creating a mold to mass-produce a product, it may want to ensure that its designs are effective and meet quality expectations. To do this, they may 3D-print dozens of iterations, or prototypes, until the desired quality is achieved, as shown in *Figure 10.2*:

Figure 10.2: Using 3D printing as a rapid-prototyping technique

At one point, prototyping was typically completed entirely by hand, which took an extensive amount of time. The introduction of computer-aided design software and 3D printers has reduced prototyping time and costs, while increasing efficiency. As shown in *Figure 10.2*, companies and now makers alike can create endless prototypes and samples to test their ideas using 3D printers, before looking to invest in mass-producing a part using another production method. These prototypes can be made on a different scale, use a different material, and perhaps may not even function exactly like the final product, but they can be developed at a rate and quality that was not previously possible.

But modern 3D printers have now become so quick, so versatile, and of such high detail that they are starting to be used to create final products more and more. This has been propelled by the *maker movement*, consisting of creators and artisans making their own products and selling them at craft shows, in online stores, or on sites such as *Etsy*. **Print farms** typically consist of three or more 3D printers, grouped to create small-to-large-scale production facilities, like the one shown in *Figure 10.3*:

Figure 10.3: LulzBot's 3D printer farm (image credit: www.lulzbot.com)

Large-scale print farms like the one used by *LulzBot* shown in *Figure 10.3* manufacture hundreds of parts a day, using about 200 3D printers. And while LulzBot uses its own 3D printers to make parts for other 3D printers, print farms are growing in popularity in both industrial and hobby settings to make a variety of products. It is now increasingly common to find products available to purchase made using 3D printers, from ornaments and jewelry, to prosthetics, sneakers, or even another 3D printer.

And now that we've covered what 3D printing is and how it is used, let's dive deeper into how it works.

How does 3D printing work?

You may not have 3D-printed something before, but I am willing to bet you've printed something on paper using a boring, old-fashioned printer at least once. While this printing method doesn't get as much hype as it used to, we can use it to make some connections between the different processes.

In order to print something (on paper), we need to complete a few steps, as shown in *Figure 10.4*:

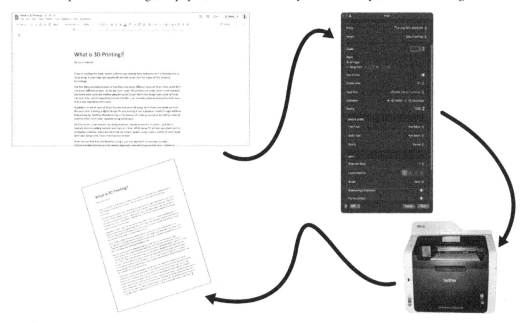

Figure 10.4: An overview of the steps to print a document using a typical desktop printer

As illustrated in *Figure 10.4*, I need a design before I can print anything. Sure, it might be a simple 2D design, perhaps a letter, a poster, or even this book. Regardless, it needs to be formatted so that it will fit on the paper loaded into my printer. After I have my design ready, I can press *print* to send it off to the machine.

But before the machine receives it, the document is processed through a print driver. This driver reformats my design, adjusts the margins and color profiles to match what the printer has available, and even checks to see whether the printer has the right-sized paper loaded all based on my design. If necessary, adjustments can be made, such as selecting grayscale rather than color or resizing the document to fit on the paper available.

After all of the formatting is complete, the printer will read the information sent from the driver to produce a document we can actually hold in our hands, based on our original design file. And the printer will do this to *its best ability*. What I mean by this is if I send a high-quality color photo to a lower-quality printer, it might not look as good on the paper as it did on the screen. It might not even be printed in full color if the printer can't do that. Additionally, our design might need to be scaled to fit on the paper available if we created a document with margins bigger than what the printer can actually handle.

I'm covering 2D printing here because 3D printing works in a very similar way. We have learned how to make complex and uniquely designed 3D models in Tinkercad, but there are additional steps and considerations involved in creating these models effectively with a 3D printer, as shown in *Figure 10.5*:

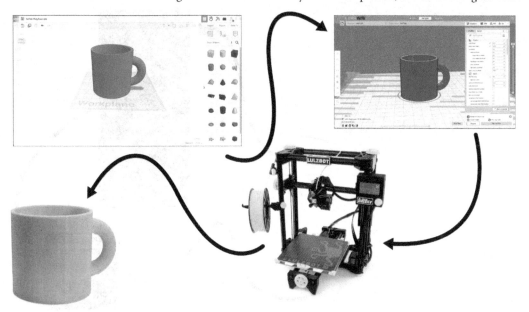

Figure 10.5: An overview of the steps to 3D-print a model using a typical 3D printer

Like traditional printing, 3D printing starts with a design, but of course, this design needs to be a 3D file rather than a 2D document. As we've learned, we can create these 3D designs in CAD programs such as Tinkercad, as well as find models on websites such as *Thingiverse* or *Cults3D*. However, as illustrated in *Figure 10.5*, we can't simply press print and have a 3D printer create our 3D models; instead, we need to send them to a different type of program called computer-aided manufacturing first.

Computer-Aided Manufacturing (CAM) software works kind of like how a print driver works in traditional printing. In CAM, we can ensure that our model fits on our 3D printer and choose to scale or rotate it as needed. We can't change the design in CAM, which would have to be done in our CAD program, but we can adjust the position or orientation to optimize printing. If we have the space, we can even import more than one model and 3D-print multiple parts at a time. We also use CAM to select the resolution of our print, as well as ensure that the print settings correspond to our printer and the selected material used for manufacturing.

While there are some generic CAM programs, such as *Cura by Ultimaker*, many 3D printers use their own unique CAM software and settings. All 3D printers need this crucial step, regardless of the type of printer that is used. After importing our 3D model and adjusting the settings in CAM, the CAM software "slices" the model layer by layer to create a coded file that instructs the 3D printer how to make our part, one movement at a time. This is why CAM programs for 3D printers are sometimes called **slicers**, and the coded file that is created and sent to a 3D printer is called a **Gcode** file. While it is important to have a general understanding of this process here, we will dive deeper into CAM software and the steps for production later in *Chapter 14*.

Like traditional printing, a 3D printer will attempt to recreate your 3D model to *its best ability*. For example, if you design something in blue but green material is loaded in the printer, your part will be manufactured in green. If you make a multicolor design but your 3D printer can only print one color at a time, then it will be made in a single color. If you attempt to 3D-print an intricate piece of jewelry on a lower-quality 3D printer, it won't look nearly as well in the real world as it does on the computer screen, and so on.

Because 3D printers manufacture parts one layer at a time, we typically refer to the layer size when measuring resolution and quality, as shown in *Figure 10.6*:

Figure 10.6: Comparing the same model 3D-printed at different resolutions

The smaller the layer height, the smoother the part, and therefore, the higher detail you can obtain, as shown in *Figure 10.6*. This model of a little boat is called *3DBenchy*, and it is a challenging model, typically used to "bench test" the quality of a 3D printer. As you can see, the different boats shown in *Figure 10.6* get increasingly smoother and of higher detail as we progress toward the right side of the image. To do this, higher-quality settings were chosen in CAM software, or a higher-quality 3D printer was used to create a more desirable finish.

However, a perfectly smooth model and small layers are not always necessary to find success. If you are prototyping, you might be more concerned with speed, and larger layers allow you to print more quickly. You might also be looking to create a strong part, and thicker layers may also offer more rigidity. We'll look into these different settings and considerations to 3D-print our parts further in *Chapter 14* when we get hands-on with manufacturing our designs.

Now that we know how 3D printing works, we will dive deeper into the different production techniques used from 3D printer to 3D printer.

Comparing 3D printing techniques

When we refer to 3D printing techniques, we refer to the mechanical process that is used to create a model, layer by layer, through an additive approach. You may be surprised to learn that there are many different types of 3D printers out there, each using a different type of material and method to create a 3D part. We won't be learning about all the available options in this book; instead, we will focus on the ones that are most widely used in both hobby and industrial settings.

Fused Filament Fabrication

Fused Filament Fabrication (**FFF**) is one of the most versatile and commonly used 3D printing production methods. The way this type of 3D printer works is by melting a roll of material, called a **filament**, while pushing that material through a nozzle, as shown in *Figure 10.7*:

Figure 10.7: An FFF-style 3D printer in the process of 3D printing

To understand how this works, think of a hot glue gun. Like glue sticks being fed through a hot glue gun's nozzle, the filament is fed through the nozzle of the 3D printer in a part called an **extruder**, which can be seen in *Figure 10.7*. The diameter of the nozzle impacts the detail and speed possible, and some 3D printers, such as the *LulzBot* shown in *Figure 10.7*, can be equipped with different nozzles and extruders for different types of projects, such as making finely detailed parts or printing something very quickly. Many FFF-style 3D printers also often have heated platforms to print onto, called **print beds**. Heated beds make adhesion more effective, which allows for a wider range of filaments to be used.

FFF 3D printers offer a few benefits that make them very popular, as highlighted by the dual-extruder FFF printer shown in *Figure 10.8*:

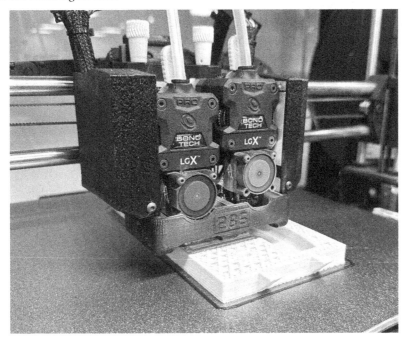

Figure 10.8: A dual-extruder FFF 3D printer in the process of 3D printing (image credit: `www.lulzbot.com`)

The 3D printer shown in *Figure 10.8* can print with two different extruders for a single design. This means you can make a multi-color print, as well as make a model with two different types of materials, such as rigid plastic and flexible plastic, to make a unique part. There are FFF 3D printers that can print with two, four, or even up to eight different materials at a time to create really complex creations. Other benefits of using a FFF 3D printer are as follows:

- FFF printers are usually more affordable to purchase and operate
- FFF printers are versatile in the types of materials that can be used
- FFF 3D printers come in all different sizes, from desktop 3D printers to 3D printers that are even larger than a kitchen refrigerator

However it is important to note that FFF printers operate by using the same fundamental production technique as **Fused Deposition Modeling** (**FDM**) 3D printers, which we will discuss next.

Fused Deposition Modeling

Both FFF and FDM 3D printers use an extruder and nozzle to melt rolls of filament that are fused together in layers, but the key difference between these two classifications is that FDM printers also incorporate a sealed and heated printing environment, as shown in *Figure 10.9*:

Figure 10.9: An FDM-style 3D printer

Compared to FFF, FDM printers are typically found in more industrial settings rather than hobby or education settings because the enclosed chamber allows for a wider range of materials to be used, and offers a more controlled and precise printing environment. However, FDM extrusion-type 3D printers, like the one shown in *Figure 10.9*, are not the only examples of a widely used additive manufacturing technique in industry. We can compare these types of printers to another common technique, called **vat photopolymerization**.

Vat photopolymerization

Vat photopolymerization is commonly referred to as **resin printing**, and while the products of these techniques are fundamentally the same as FFF or FDM, the way resin printers work is vastly different. There are no extruders or nozzles, but rather, a light source. And instead of using rolls of filament, a liquid resin is used. While the types of light sources and resins can vary from printer to printer, all vat photopolymerization 3D printers use their light source to convert liquid resin into a rigid object, as shown in *Figure 10.10*:

Figure 10.10: A close-up of a part being printed using the SLA technique (image credit: www.creative-tools.com)

As shown in *Figure 10.10*, a part 3D-printed through vat photopolymerization is still created one layer at a time, but the process differs vastly from FFF- or FDM-type machines. The manufactured part is pulled vertically out of a tray of liquid resin as the printing platform, or build plate, raises each layer. The light source is typically housed below the tray of resin, and this printing area is fully enclosed to prevent contamination or stray light rays, as shown in *Figure 10.11*:

Figure 10.11: An SLA resin 3D print farm (image credit: formlabs.com/industries/dentistry/)

Not all resin printers are the same, like the ones shown in *Figure 10.11*. When comparing different resin printers, you may want to consider the types of resins that can be used as well as the type of light source, both of which may impact the quality and production capabilities. The more common **Stereolithography** (**SLA**)-style printers like the ones shown in *Figure 10.11* use a UV light source to cure the liquid resin.

Stereolithography

One of the key benefits of this type of production technique is that exceptionally high detail can be achieved to create the most intricate of parts, as shown in *Figure 10.12*:

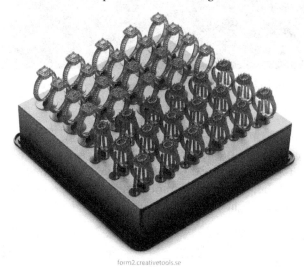

Figure 10.12: 3D printing ring investment models produced with a wax resin (image credit: www.creativetools.se)

As shown in *Figure 10.12*, models manufactured with resin printers can be highly detailed and made from a range of material types, including wax. Thanks to its detail, versatility, and speed, resin 3D printing is widely used for industrial applications in both prototyping and production processes. Like FFF and FDM, there are also a lot of different types of resins that can be used, from rigid plastics to flexible ones, or even medical-grade materials. Compared to extrusion-type 3D printing, resin-type printers can typically achieve higher resolutions, but they also often have a smaller printing volume and typically cost more to operate.

In addition to choosing between different printing techniques, different postproduction techniques need to be considered, from sanding your prints to putting them into curing chambers. These strategies and methods will vary based on the type of 3D printer you use, as well as the type of material you select.

Choosing the right material

As you consider the design or part you would like to create, you now know that you can choose from a range of 3D printing production techniques to do so. Choosing different types of 3D printers can change the complexity of the type of product you create, including the material, size, or quality, like the product shown in *Figure 10.13*:

Figure 10.13: A 3D-printed coffee cup made from 3D-printed plastic

The mug shown in *Figure 10.13* was made using a FFF 3D printer. This was printed in a plastic that is vibrant in color and very affordable, but not one that could withstand high temperatures or one that is food-safe. As such, this mug would be great to hold pens and pencils on a desk, but maybe not to actually drink coffee out of. If I wanted to drink from this, perhaps I should have chosen a 3D printer that could print clay to make this part, like the one shown in *Figure 10.14*:

Figure 10.14: A vase 3D-printed with clay

While the models shown in *Figure 10.13* and *Figure 10.14* have been created using two different design files, the biggest difference is the material used to make each model. Two different types of 3D printers were used to make these parts, which provided different options to work with, such as choosing between quality, speed, and material. Having an effective design is one thing, but manufacturing it so that it can be used effectively is something else to consider entirely.

Fortunately, there are many material choices to pick from that work with a wide range of 3D printers, so you don't necessarily need a different type of printer for different project circumstances.

Printing with polylactic acid

Polylactic Acid (PLA) is a fantastic material choice for many different projects, as it is non-toxic, non-allergenic, and is a more environmentally friendly choice compared to other types of materials. It is also available in many different colors, as shown in *Figure 10.15*:

Figure 10.15: The 3DBenchy test model 3D-printed in different color materials

In addition to its vast color selection, PLA is also an easier material to work with as it can be printed at lower temperatures, and even without a heated bed (for FFF and FDM applications). PLA is also an affordable material and one that can easily be sanded, painted, and glued through postproduction.

When using an extrusion-type printer, you may also find different types of PLA filament that can be used to create unique parts and prototypes effectively, as shown in *Figure 10.16*:

Figure 10.16: A 3D-printed trophy using a silk and translucent PLA filament

A metallic silk PLA was used to make the base of the trophy shown in *Figure 10.16*, while a translucent PLA was used for the top part, which allows light to shine through. After printing the two parts separately on single extruder FFF printers, they were glued together to create this final design, which is something we'll look at more later in *Chapter 15*. Different types of PLA as well as PLA from different manufacturers can not only change the appearance of a part but also its properties. I have found some PLA to be more brittle than others, and some to be far more flexible or temperature-resistant. You can even find PLA that is conductive, as shown in *Figure 10.17*:

Figure 10.17: A 3D-printed flashlight (image credit: www.lulzbot.com)

Conductive PLA, such as the one used in the prototype flashlight shown in *Figure 10.17*, allows for current to flow through it when connected to a power source, such as a battery. When combined with normal PLA, which is an insulator, you can create unique circuits and prototypes, and even fully 3D-printed flashlights, as shown in this example!

Other commonly used materials

There are other commonly used types of printing filaments and resins that offer different properties and production capabilities when compared to PLA. **Polyethylene terephthalate glycol (PETG)** is more durable than PLA, allowing you to make stronger parts without needing a different type of 3D printer. **Acrylonitrile butadiene styrene (ABS)** is also commonly used for heat-resistant or industrial applications, although it is also one that needs special care and preparation to ensure health and safety due to its fumes.

Flexible materials, such as **Thermoplastic polyurethane (TPU)**, are another one of my favorite printing materials to use. These materials allow you to make rubber-like parts, as shown in *Figure 10.18*:

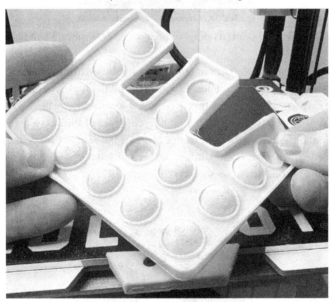

Figure 10.18: A pop-it 3D-printed using a TPU filament and an FFF printer

The pop-it shown in *Figure 10.18* isn't quite as flexible as a silicon one you can buy off the shelf, but it offers a similar flexibility and experience, despite being made using an additive method rather than a formative one. I find TPU or nylon materials to be stronger than the rigid ones we've discussed thus far, allowing for durable and flexible parts to be created. There are many different TPU filaments available, as well as different flexible resins, but not every 3D printer can print rigid and flexible filaments effectively.

There are so many more options to choose from in addition to the ones listed in this chapter, such as metal-infused or wood-infused materials, or even material made from recycled fishing nets! The first step is to consider your design, how it looks, and how it is intended to function. From there, you should then choose the 3D printer and the material that will support your design, time constraints, and budget limitations most effectively.

Summary

In this chapter, we looked at one of the most exciting manufacturing tools of today, 3D printers. As discussed, 3D printers come in all shapes and sizes and can turn rolls or bottles of material into a solid part, based on your 3D designs.

3D printers also come in many different types, including FFF or FDM printers that melt rolls of filament that are bonded, layer by layer, to create parts. Or resin-style printers that use a light source to convert liquid resin into a solid models. Each type of 3D printer has its own set of benefits and drawbacks, and it's up to us to choose the best type of printer and printing material, based on our needs.

Most beginners find success and ease when working with FFF-style printers, while FDM-style printers offer similar flexibility to FFF printers but with greater detail, which makes them suitable for more industrial settings. When striving for exceptionally high detail or making intricate models, using a vat photopolymerization-type printer is the optimal choice, despite the higher operation costs.

While we looked at the most common types of 3D printers and materials in this chapter, it's important to note that there are many more that we did not have the time to cover here. I recommend you take a bit of time to look through some of the other types of 3D printers out there, such as **Selective Laser Sintering** (**SLS**) printers, which use a laser for a range of materials and applications, and **Direct Metal Laser Sintering** (**DMLS**) printers, which can make parts using metal.

In the meantime, turn the page to dive back into Tinkercad, as we look at general strategies to create effective models for 3D printing using some of the new topics and concepts we covered in this chapter!

# 11
General Strategies for Creating Effective Models for 3D Printing

In this chapter, we will build on the topics introduced in *Chapter 10* as we look towards designing 3D models in Tinkercad which can be manufactured through 3D printing effectively. As discussed in *Chapter 10*, there are many different types of 3D printers and materials available, each with its own unique benefits and drawbacks. As designers, we can create models that take advantage of the benefits of the resources available, while also looking to reduce the negative impact of any drawbacks.

We'll be looking at strategies and skills in effective 3D modeling for 3D printing production through the following topics throughout this chapter:

- Avoiding overhangs
- Creating segments, fillets, and chamfers
- Designing for the first layer
- Optimizing the build plate

By the end of this chapter, you will see how the strategies and techniques introduced in *Chapter 10* can be supported by the skills gained in this chapter, in addition to referencing many of the introductory skills gained earlier this book. While there is never one single approach for finding success in every situation, we will be learning skills and best practices that can universally allow us to find greater success as we continue to grow in our abilities for 3D modeling and 3D printing.

Technical requirements

We will be again returning to Tinkercad in this chapter as we learn to design with some of the newly introduced concepts for 3D printing production introduced in the last chapter. As a reminder, we can access Tinkercad through a web-based design application at www.tinkercad.com.

You can also access Tinkercad via an app depending on the device you are using, such as an iPad or other tablet device. As discussed in *Chapter 2*, I recommend you utilize a mouse to aid in your design work in Tinkercad's 3D editor as it may offer an additional level of control and ease of use.

An editable model of the examples shown throughout this chapter can be accessed on Tinkercad at `https://www.tinkercad.com/things/3nyCa72nWhX-overhang-examples-from-chapter-11`.

Avoiding overhangs

As discussed in *Chapter 10*, 3D printers create models by adding material in layers. This of course allows us to make geometrically unique parts, but it also introduces an inherent drawback that we need to work around. Let's learn about this drawback more in the upcoming sections.

What is an overhang?

As an example, let's look at the Tinkercad model shown in *Figure 11.1*:

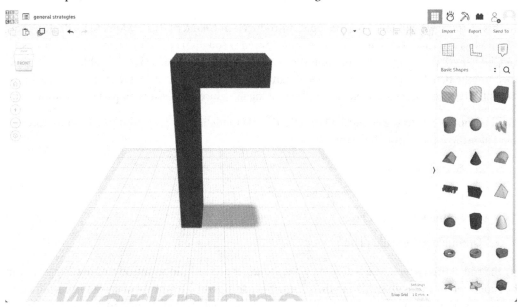

Figure 11.1: An example Tinkercad model with an overhang

The 3D model shown in *Figure 11.1*, which resembles an upside-down *L*, has what is referred to as an **overhang**. As initially discussed in *Chapter 5*, overhangs are sections of a part that are unsupported, meaning they are hanging out with nothing underneath them, including no other structural parts of the model. As a 3D printer attempts to make this part, the overhang will begin to fail or cave in as it is just floating in midair and gravity will naturally take over, as shown in *Figure 11.2*:

Figure 11.2: Printing a part with an overhang on a FFF 3D printer

As shown in *Figure 11.2*, the overhang droops and splits as the length of the overhang increases during production because there is nothing to support these overhanging layers. This could be a catastrophic problem if not addressed, though it is important to note that the overhang shown in *Figures 11.1* and *11.2* is quite extreme as it is a 90-degree angle. The smaller the angle, the easier the printing process will be as the part can support itself, as shown in *Figure 11.3*:

Figure 11.3: Comparing overhang angles

As shown in *Figure 11.3*, a smaller angle will allow for layers to build on one another, which naturally supports the overhang. Depending on the type of 3D printer you are using *(FFF, FDM, SLA, etc.)* your printer may be able to handle steeper overhangs with greater success. We can also adjust the print speed and cooling settings in our CAM software to enable greater overhang performance too.

How can we avoid an overhang?

But we can also mitigate this challenge by simply changing the orientation of our part to avoid an overhang altogether. I may want to use the part shown in *Figure 11.1* as an upside down *L*, but I can print it in a different orientation for better performance, as shown in *Figure 11.4*:

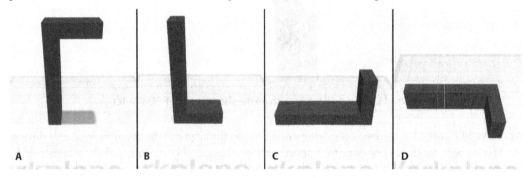

Figure 11.4: Changing the part orientation to remove overhangs

As seen in *Figure 11.4*, the original part is shown in view *A* and has a steep 90-degree overhang. But when rotated, as shown in views *B*, *C*, or *D*, the part can be manufactured with no overhang, which in turn will make a higher-quality and stronger part. As we design our models, it is first important to consider how we intend to use them. But we then must consider the nature of 3D printing techniques and then adjust our models to not only perform as intended, but to also be manufactured in the most effective way.

But sometimes overhangs are unavoidable and necessary to successfully create an effective design. In this case, we can utilize a manufacturing technique called **support material**. Support material can be automatically generated in **CAM** software before a *Gcode* file is created to 3D print our part, as discussed in *Chapter 10*. These supports will fill the cavities under overhangs using a minimal amount of material that can then be removed after manufacturing the part, as shown in *Figure 11.5*:

Figure 11.5: A part printed with support material under an overhang

The zig-zag support material shown being removed from the part in *Figure 11.5* adds a little bit of time to our printing process, as well as a bit of time in the post-production process, but it allows for overhangs to be printed more effectively. Sometimes sanding might be needed for a higher-quality finish as supports don't always break away smoothly, but they also allow us to make geometrically challenging models without needing to sacrifice our design. There are also many different methods for using support material, such as printing **tree supports** as shown in *Figure 11.6*:

Figure 11.6: 3D printing a part with tree supports

Like the more traditional zig-zag support method shown in *Figure 11.5*, tree supports can be automatically generated in the CAM software if your 3D printer supports it. Tree supports follow a more organic approach to supporting overhangs as shown in *Figure 11.6*, by wrapping up and around your design with support branches. This typically cuts back on wasted material, and also is usually easier to remove, which produces a higher-quality finish with less effort required in the post-production stages.

And if you find yourself using an FFF or FDM 3D printer that can print with more than one material at a time, you can then use dissolvable support material as shown in *Figure 11.7*:

Figure 11.7: A 3D printed part manufactured with dissolvable support material (image credit: www.lulzbot.com)

As shown in *Figure 11.7*, dissolvable support material can be printed around your model to support all overhangs when using a multi-extruder 3D printer. After printing the part, it can then be soaked in a water-based bath, which dissolves all support material leaving a high-quality part with minimal effort in the post-production phases. This technique is widely used in industrial applications, and usually requires a more advanced 3D printer.

Resin-type 3D printers using the SLA printing process outlined in *Chapter 10* also have a different tolerance towards printable overhangs, and typically require more supports to be added compared to extrusion-type printing techniques such as FFF and FDM. But because resin printers are often of a higher resolution, these supports can be printed in very fine detail, almost like a wireframe around a model as shown in *Figure 11.8*:

Figure 11.8: A 3D model printed with an SLA printer using supports for overhangs (image credit: formlabs.com/3d-printers/)

Like the zig-zag and tree supports discussed earlier, the framework of supports shown in *Figure 11.8* would be automatically generated in your printer's CAM software. You could then peel or cut away the supports from where they connect to the model, then clean up the model through postprocessing steps including sanding or curing in a UV bath. But again, overhangs can often be avoided or reduced by simply reorienting our parts or designing with 3D printing techniques in mind. Using support material may allow us to create complex parts, but it also adds a layer of complexity, time, and extra resources needed as well. The best strategy for successful production is to design cleverly to reduce the need for supports, then utilize supports when and only where necessary.

We'll be looking at how support material can be enabled automatically in CAM software later in *Chapter 14*, but let's now look at how we can make the quality of our models even better by smoothing out their corners and increasing their evenness.

Creating segments, fillets, and chamfers

We can work to increase the quality and rigidity of our models by increasing the segments of our Tinkercad designs, or add fillets and chamfers in our overhangs to design our 3D models more effectively too.

Adjusting segments

As initially introduced in *Chapter 4*, many of the *basic shapes* in Tinkercad's **Shape Library** have parameters to adjust the outer surface, such as the **Cylinder** shape shown in *Figure 11.9*:

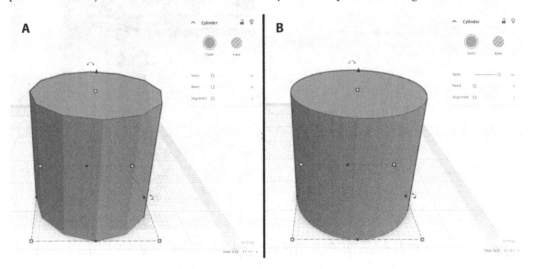

Figure 11.9: Adjusting the side parameter of a cylinder shape

The parameter we can adjust varies a bit from shape to shape. Sometimes this is called **Segments**, as when creating a beveled edge; sometimes it is called **Steps**, as when working with the box shape; in the case of the cylinder, shown in *Figure 11.9*, it is called **Sides**. By increasing the number of **Sides** in the shape parameters for the cylinders shown in *Figure 11.9*, we can make it appear smoother. This is because the cylinder shown in view **B** has a higher number of sides, so its outer surface is smoother than the one shown in view **A**. This change in sides is also noticeable when 3D printing these two parts, as seen in *Figure 11.10*:

Figure 11.10: 3D printing cylinders with varying sides

As seen in *Figure 11.10*, the two cylinders are noticeably different just as they are in the Tinkercad designs shown in *Figure 11.9*. While this example shows an outer surface impacted by side count, the same effect would take place on inner surfaces or overhangs as well. This can not only impact the final appearance of your models, but their performance too. Typically, smoother parts are of a higher quality when 3D printing and often create a more desirable product.

Another modeling strategy that can increase the performance of our final product is adding a **fillet** or **chamfer** to the edges of our parts. Let's look at this in the following section.

Creating fillets and chamfers

Fillets and chamfers have a similar use as sides and segments, but also have a different appearance as shown in *Figure 11.11*:

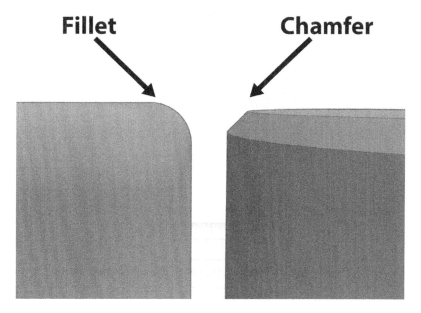

Figure 11.11: Comparing the difference between a fillet and a chamfer

Even though there isn't a specific fillet or chamfer tool in Tinkercad, we can still make these types of features in our designs using the tools available. As seen in *Figure 11.11*, fillets are rounded edges while chamfers are angled ones. The intention of these strategies is usually the same: create a higher-quality finish on a model, or increase model strength. You can choose which approach you want based on personal preference, or the options available in Tinkercad's shape parameters.

For example, the **Box** shape allows us to create a fillet in the shape parameters window by adjusting the **Radius** parameter, as shown in *Figure 11.12*:

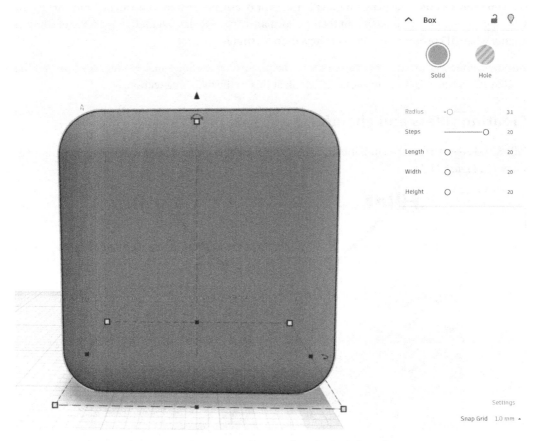

Figure 11.12: Creating a fillet on the edges of a box shape

The **Cylinder** shape allows us to create a chamfer instead of a fillet by adjusting the **Bevel** parameter, as shown in *Figure 11.13*:

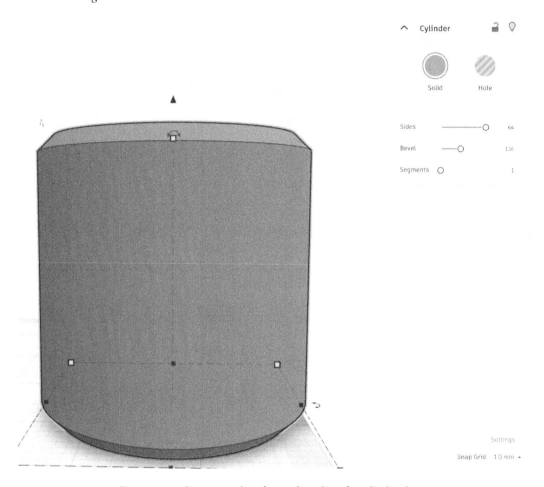

Figure 11.13: Creating a chamfer on the edge of a cylinder shape

Looking at *Figures 11.12* and *11.13*, our design decisions may sometimes be made for us based on the shapes and options available in Tinkercad, unless we look to import additional shapes, as discussed in *Chapter 8*.

Using CSG to create fillets and chamfers

We can also utilize the concepts of CSG to create fillets and chamfers to support our overhangs, which is not only a strategy for increasing the final finish of a part, but also increasing its overall strength. This can be done through the following steps:

1. To start, we can create a model with an overhang as shown in *Figure 11.14*:

Figure 11.14: Starting with a model that has an overhang

2. If we wanted to add a chamfer to support this overhang, we could then add a **Wedge** shape in the corner of the overhang as shown in *Figure 11:15*:

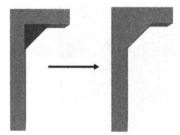

Figure 11.15: Creating a fillet to support an overhang

3. Alternatively, we could utilize the **Box** and **Cylinder** shapes to create a fillet instead. First, select a **Box** shape and a **Cylinder** shape and drag them on to your Workplane as shown in *Figure 11.16*:

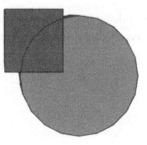

Figure 11.16: Selecting shapes to make a fillet through CSG

4. Next, turn the **Cylinder** shape into a hole shape and use it to cut the **Box** shape to create a corner-fillet shape, as shown in *Figure 11.17*:

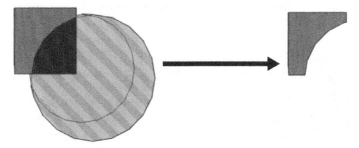

Figure 11.17: Combining shapes to make a fillet through CSG

5. Lastly, position this corner-fillet shape into the corner of our overhang to create a fillet, as shown in *Figure 11.18*:

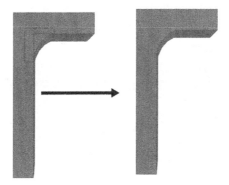

Figure 11.18: creating a fillet to support an overhang

Overhangs, segments, fillets, and chamfers are all important considerations for creating more effective models for 3D printing, and are also factors in making models with a higher-quality finish or that perform better. But the most important place to consider these concepts is without a doubt on the first layer.

Designing for the first layer

To understand what I mean by *first layer*, we must go back to the conversations on how 3D printing works as introduced in *Chapter 10*. Regardless of the type of printing technique we use, our models are designed in layers, building the next one on top of the last one. In extrusion-type printing techniques such as FFF or FDM, these layers typically build from the bottom up on a print bed, while resin-type printing such as SLA pulls a model out of a liquid layer by layer to create the model.

For each of these techniques, every subsequent layer is built upon the very first one, either up or down. And if that first layer fails, meaning that it caves in because it isn't properly supported or perhaps it slips off the print bed due to poor adhesion, the whole model is usually destroyed as a result, as seen in *Figure 11.19*:

Figure 11.19: A failed print

It is often the case that failed prints like the one shown in *Figure 11.19* occur due to faulty settings in the CAM software, or perhaps a poorly calibrated printer. But there are also design strategies that can be employed to reduce the likelihood of these failures and also increase the effectiveness of our models. Let's look at these strategies in the next section.

Utilizing surface area for better parts

Earlier in this chapter, we discussed how the different orientations of the part shown in *Figure 11.4* would be better in reducing overhangs as rotated in views **B**, **C**, or **D**. But when discussing layer adhesion and overall print performance, the orientation shown in view **C** or **D** would be the best option.

This is because the orientation in views **C** and **D** have the most surface area touching the build plate, as illustrated in *Figure 11.20*:

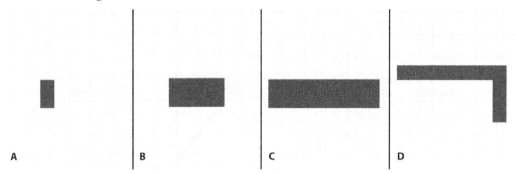

Figure 11.20: Comparing the first layer's surface area based on part orientation

The views in *Figure 11.20* match the parts shown in *Figure 11.4*, but specifically show the parts from the top and only display their cross-sections where the part contacts the print bed. As seen in *Figure 11.20*, views **C** and **D** have the most surface area touching the print bed for the first layer, which means they are the most likely to have the best adhesion during printing due to a higher amount of contact. The orientation in view **D** is also the flattest of the designs, which would make it most likely to perform the best for overhang avoidance, strength, speed, and print bed adhesion when printed through an extrusion-type process such as FFF or FDM.

As another example, let's say you are looking to print a rounded-sphere part as seen in *Figure 11.21*:

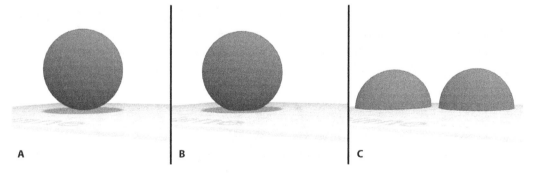

Figure 11.21: Increasing the first layer's surface area for a sphere shape

The original sphere shown in view **A** of *Figure 11.21* would be very difficult to print because such a small amount of surface area is touching the print bed for the first layer. To print this successfully, a combination of support materials would probably be needed. But if your design allows for it, perhaps you can consider adding a flat spot to the sphere as shown in view **B**, or even splitting the sphere and printing it in two parts, to then be glued together in post-production as shown in view **C**. Both options **B** and **C** would increase the likelihood of successful printing and may also create a stronger and higher-quality part at the same time.

Adding rafts and brims

Much like support material, which can be automatically generated in CAM software to aid in the event of overhangs when they can't be avoided, there are strategies for supporting first-layer adhesion that can be added to our designs as well. **Rafts** and **Brims** can be added to our models to increase the surface area contacting the print bed, as shown in *Figure 11.22*:

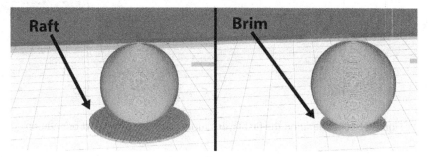

Figure 11.22: supporting first layer adhesion in Cura LulzBot Edition

The steps for adding a raft or brim to your model in 3D printing CAM software will vary a bit based on the model 3D printer you are using, but the general concept is the same. As shown in *Figure 11.22*, I can use the *Cura* slicer program for my *LulzBot* 3D printer to automatically create a raft as shown in the view on the left, or a brim as shown in the view on the right.

A raft essentially creates a removable platform that your model is printed on to ensure a large amount of surface area is used for the first few layers. This is a more traditional solution to this problem, and one that isn't used often anymore for extrusion-type 3D printers that have heated print beds, but are still frequently used for resin-type printing as resin printing platforms are quite different in design from a heated bed. For extrusion-type printing, I often use a brim to support the first layer as this is a thin perimeter that is printed around the base of my model which is easy to peel away in post-production, as seen in *Figure 11.23*:

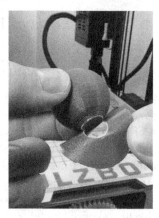

Figure 11.23: Removing the brim off of a 3D printed part using a FFF 3D printer

But like support material, rafts and brims should really only be used when we cannot design our models to effectively increase surface area contact and print bed adhesion for the first layer. When considering the print bed, another thing we can do to increase the effectiveness of our modeling techniques is to optimize the build plate in our designs.

Optimizing the build plate

Let's say that we want to 3D print multiple parts at the same time for prototyping or production. CAM programs allow you to import multiple models to prepare for 3D printing, but this can often be done effectively in Tinkercad as well.

The first thing we want to do is to set our Workplane to be the same size as our print bed using the **Workspace settings** window as shown in *Figure 11.24*:

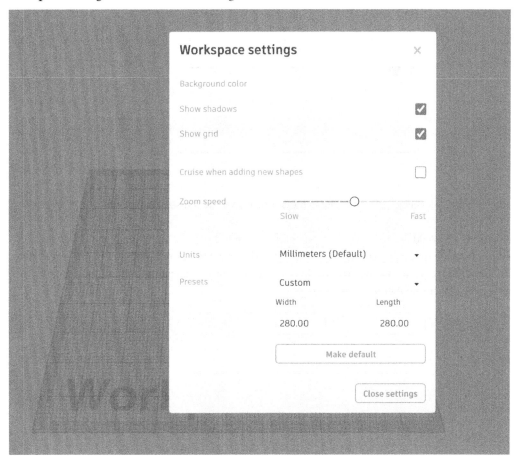

Figure 11.24: Adjusting the Workplane size in Workspace settings

As discussed in *Chapter 6*, Tinkercad provides some common 3D printer bed sizes to choose from in the **Presets** dropdown, but we can also enter a **Custom** size to match our own 3D printers as shown in *Figure 11.24*. If you would like, you can even set this to be the default using **Make default** so that each new Tinkercad document you create will have this already set for you. With the Workplane set to match my 3D printer's bed (280 mm x 280 mm), I can now get an accurate sense of how much space I have to work with, as shown in *Figure 11.25*:

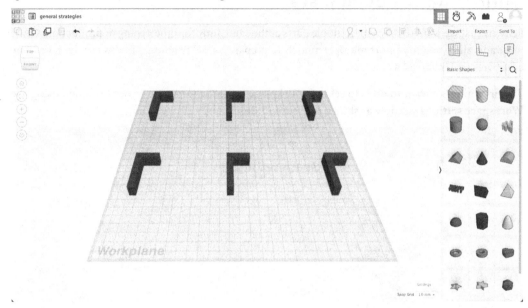

Figure 11.25: Multiple parts shown on a custom-size Workplane

Let's say that I would like to 3D print the six parts shown in *Figure 11.25*. If I were to print them spaced out like they are in the preceding figure, time would be wasted as the 3D printer has to travel quite a distance to move from part to part. I could easily reduce my overall print time just by arranging these parts to be closer together as shown in *Figure 11.26*:

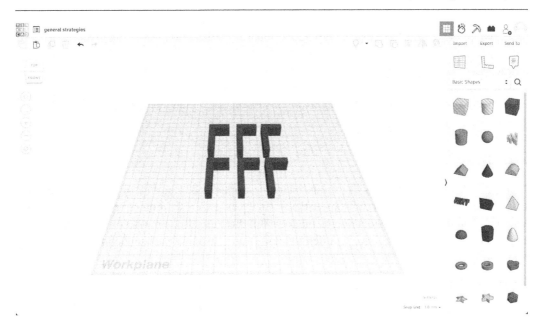

Figure 11.26: rearranging parts to reduce print time

The same parts are now positioned more closely together in *Figure 11.26*, and this will reduce my overall print time by 18 minutes (with the default settings for my printer specifically). Let's say I wanted to print more of these parts even more quickly for production. In that case, I can work to nest the parts within one another to fill my Workplane as shown in *Figure 11.27*:

Figure 11.27: Nesting parts to fill a Workplane

As shown in *Figure 11.27*, Tinkercad's design tools make rotating and positioning parts very easy to do using some of the tools we learned about earlier in this book, such as **snap grid** and **duplicate and repeat**. While this can often be done in a CAM program, you may find that organizing your parts in Tinkercad before downloading your models for production is a more effective and efficient approach, and also one that was important to cover before we dive deeper into more CAM techniques and steps for production later on.

Summary

Looking back at the topics and skills introduced in this chapter, it is important to remember that there are countless situations and circumstances that will vary as you create different models and interact with different 3D printing production techniques. But there are also common practices that can be used to increase the effectiveness of our designs, and in turn, increase the likelihood of success when 3D printing.

When possible, always look to reduce overhangs in your design by changing the part orientation, or by creating fillets and chamfers to support the overhangs during production. Fillets and chamfers may also increase the overall aesthetics of your part, along with the strength of the part after it has been printed as well.

Likewise, we should always work to increase the amount of surface area in contact with our build platform for the first layer, as a failed first layer will almost definitely result in a failed print. Changing part orientation or separating a model into multiple parts are both effective strategies to employ in our designs, and something to look out for if you find yourself sourcing existing models and parts to bring into your own creations too.

While 3D printer CAM programs such as Cura allow for supports, rafts, and brims to be automatically generated to increase the effectiveness and success of our designs, we as designers can work to create more effective models before bringing them into CAM too. This ranges from changing the design of a part, its orientation, or even arranging multiple parts to optimize the space of our 3D printer's bed. These tools will be covered in greater detail later when we move on to exporting our designs for manufacturing in *Chapter 14*.

But now that we've covered these general strategies for designing effective models for 3D printing, it's time we move into some more complex concepts such as creating multi-part models that can fit together. When you're ready, turn the page to learn how to measure and create tolerances in your complex 3D designs!

12
Creating Tolerances for Multi-Part Designs

As we move further into the third part of this book, we will continue to expand our skills for developing effective designs in Tinkercad with the intention of 3D printing our designs. As discussed earlier in *Chapters 2* and *6*, accurately dimensioning our designs is key when striving to create effective 3D models. In this chapter, we will take these concepts one step further as we determine how to adjust our dimensions in order to better support the 3D printing resources that we have available to us.

To engage with these concepts, we will be breaking topics down into the following sections:

- What are tolerances?
- How to calculate tolerances
- Adding tolerances to our dimensions

By the end of this chapter, your modeling skills will not only increase but you will also gain confidence that your models will be able to be 3D printed in a manner that is effective and functional too.

Technical requirements

Throughout this chapter, we will be discussing adjustments that need to be made based on the constraints of specific 3D printing production techniques and material choices. As discussed in *Chapter 10*, there are many different types of 3D printers available, as well as different types of materials that can be used to manufacture your designs.

This chapter will focus on general practices for the most common situations you may find yourself in, but it is important to note that identifying the 3D printing resources available to you, as well as the limitations of your available resources, is necessary to engage with the concepts introduced in this chapter at the highest level of success.

184 Creating Tolerances for Multi-Part Designs

To compare and test your own resources to the ones that are shown in this book, you can obtain the tolerance test model that will be shown later in this chapter from my Thingiverse page at `https://www.thingiverse.com/thing:6551967`.

Additionally, the example models shown in this chapter can be obtained through the following links:

- `https://www.tinkercad.com/things/0Xs5560yjpZ-fit-tester-from-chapter-12`
- `https://www.tinkercad.com/things/8GLQzQOjUii-marker-organizer-from-chapter-12`

What are tolerances?

In all types of manufacturing, designers must consider **tolerances.** Tolerances describe an acceptable deviation in the measurements of a design to ensure that it still looks or functions as intended. Few manufacturing processes are 100% accurate. It is often the case that the cost of manufacturing comes down to an acceptable tolerance range, meaning that the more accurate you want your parts to be, the more it is going to cost you.

3D printing production techniques are no different, and in this chapter we will be looking at the tolerances that we can expect when striving to 3D print our Tinkercad designs. In general, 3D printed parts can shrink or expand during the manufacturing stages, which means that our manufactured models may actually be different in size than they are in our Tinkercad designs. For example, let's look at the model shown in *Figure 12.1*:

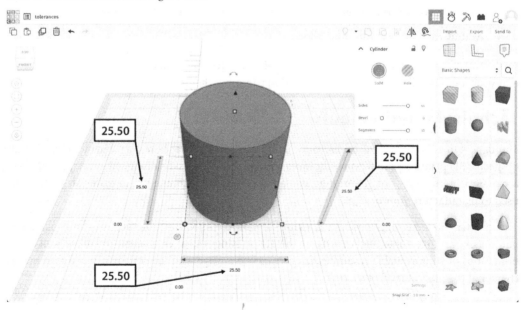

Figure 12.1: A cylinder in Tinkercad with measurements shown

As we can see, the cylinder shape shown in *Figure 12.1* is 25.5 x 25.5 x 25.5 mm in length, width, and height. Now, we would expect that after 3D printing this model, the measurements would be the same. To compare, let's look at the 3D printed part in *Figure 12.2*:

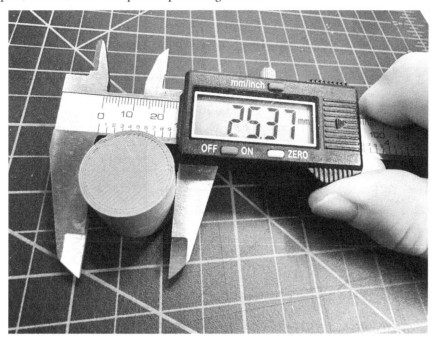

Figure 12.2: Measuring a 3D printed cylinder with digital calipers

The part being measured in *Figure 12.2* is the same cylinder shown in *Figure 12.1* after being manufactured using a **Fused Filament Fabrication** (**FFF**) 3D printer with PLA filament. By using digital calipers, we can accurately measure the size of this part, and we can clearly see that it does not quite match the original model shown earlier. It is approximately 0.13 mm smaller, in fact, and this difference in size is what we refer to as our tolerance.

However, tolerances can vary based on the type of 3D printer you are using, the quality of the printer, the printing material, and even your printing environment (the room that you're 3D printing in). It can sometimes be difficult to determine the tolerances that you should be considering in your design, which is why we are now going to take a closer look at calculating them.

How to calculate tolerances

As mentioned, tolerances can vary based on several factors, so we will be looking at general strategies for the most common instances as we consider how to calculate them for our designs.

Determining tolerance based on printer type

The first thing we need to do is determine the general tolerance of our 3D printers as this will have an impact on the tolerances to expect for our designs. In *Chapter 10*, we looked at three of the most common types of 3D printing techniques, **FFF**, **Fused Deposition Modeling** (**FDM**), and **Stereolithography** (**SLA**). We will be comparing the tolerances between these techniques, as shown in *Figure 12.3*:

3D printer	General tolerances for the lower limits		
FFF	± 0.5%	± 0.5 mm	± 0.02 in
FDM	± 0.3%	± 0.3 mm	± 0.012 in
SLA	± 0.2%	± 0.1 mm	± 0.0039 in

Figure 12.3: Comparing tolerances between common 3D printing techniques

FFF is the most affordable extrusion-type 3D printing technique, but it is also on the lower end in terms of accuracy. In general, FFF printers have a tolerance of +/- 0.5%, or about 0.5 mm or 0.02 in. This means that a model 3D printed with a FFF 3D printer, such as the one shown in *Figure 12.2*, can vary in size by this much under general circumstances.

FDM works similarly to how FFF printers work, but typically with higher tolerances. As shown in *Figure 12.3*, FDM printers can have a tolerance of +/- 0.3%, or about 0.3 mm or 0.012 in. This increase in accuracy is a reason why FDM printers are a bit more complex, and typically more expensive than FFF printers.

SLA is the common resin-type 3D printing technique that we learned about earlier. In general, this printing method is far more accurate than extrusion-type methods such as FFF or FDM. As such, we can typically expect tolerances of +/- 0.2%, or about 0.1 mm or 0.0039 in.

However, it is again important to note that these are general tolerances and that they are tolerances on the lower limits. There are certainly printers that can exceed these tolerances within each of these categories, and there are also many other factors that may impact the accuracy of your parts, as we will discuss later on in this chapter. It's also safe to say that as time goes on and this technology continues to evolve, these tolerances will only get tighter and tighter.

Determining tolerance based on material choice

Another aspect of determining the accuracy of our parts is material choice. In *Chapter 10*, we looked at some of the common types of materials used in both extrusion-type and resin-type printing techniques. Each of these materials has its own tolerances for dimension accuracy, as shown in *Figure 12.4*:

Material	General dimensional accuracies	
PLA	± 0.02 mm	± 0.00079 in
ABS	± 0.05 mm	± 0.002 in
PETG	± 0.05 mm	± 0.002 in

Figure 12.4: Comparing tolerances between common 3D printer materials

The table in *Figure 12.4* shows the dimensional accuracy of the three rigid materials that we looked at in *Chapter 10*, which were PLA, ABS, and PETG. As PLA prints at lower temperatures than ABS and PETG, it has a higher dimensional accuracy. This means that parts printed with PLA are more likely to be closer to the dimensions set in your model as PLA filament will vary by only 0.02 mm under general circumstances. ABS and PETG, on the other hand, print at higher temperatures, which means that their products are more prone to shrinking or warping. This means that the dimensional accuracy of our parts made with these materials may vary by about 0.05 mm.

We also discussed flexible materials such as TPU in *Chapter 10*, but I've intentionally excluded this type of material from the table in *Figure 12.4*. This is because this material can vary far more from brand to brand than rigid materials do. It can also vary in dimensional accuracy based on its nature as a flexible material. In general, the more flexible the material and/or the higher the temperature it prints at is, the less dimensionally accurate it will be.

Additional factors that may determine accuracy

By understanding the tolerances for both our type of 3D printer and the material that we are using, we can design our models to be manufactured at a higher accuracy. But before we move into incorporating these tolerances into our Tinkercad designs, let's identify a few other factors that may also contribute to the accuracy of our 3D-printed parts:

- **Calibration**: This is a big one as it can drastically change the effectiveness of your 3D models. If your 3D printer is misaligned or has loose belts or parts, the accuracy and consistency of your printing will decrease.

- **Printing environment**: This impacts FFF printers more than FDM or SLA ones, as the latter types are enclosed processes, but the room your printer is in may also impact your printing accuracy. If your printer is in a cold room or has a draft blowing on the print bed, this might cause your parts to shrink or shift during production, reducing your accuracy.
- **Age of material**: All printing materials, whether they be rolls of filament or bottles of resin, have a shelf life. As you get closer to the end of this, your accuracy will begin to decrease and you may even damage your 3D printer by using expired material.
- **Software settings**: This is another big one, and one we will look at in more detail in *Chapter 14*. In the CAM stages of preparing your models to be 3D printed, choosing the optimal settings for your design, printer, and material, including supports or print speeds, will impact the accuracy of your parts.

It can often be difficult to plan for all these factors, but fortunately, many of them are not directly impacted by our design. It is first important to design a model that can be manufactured accurately using the techniques we've discussed as well as the tolerances for your printer and material, then ensure that you support your design with proper setup and implementation of production techniques. The most common mistake we can make is to not consider tolerances at all or to not add enough. Let's move on to adding tolerances into our Tinkercad designs so that we can see how to avoid making these mistakes.

Adding tolerances to our designs

Now that we've learned about general tolerances for common printing techniques and material choices, we can compensate for these tolerances in our 3D models. This is an important step not only to create accurate models but also when creating multi-part models that must fit together. Let's look at the considerations for such instances.

Types of fit

When creating successful multi-part models, the first step is to consider the type of fit we want between our parts, as shown in *Figure 12.5*:

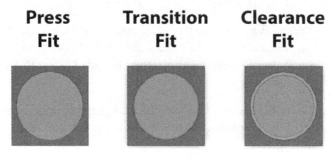

Figure 12.5: Comparing the types of fits between parts

The models shown in *Figure 12.5* compare the three general types of fits when creating a multi-part design that has two or more parts fit together, such as the hole and peg shown in these images. A **press fit** is the tightest fit with the tightest tolerances. This type of fit has two parts assembled in such a way that they are not intended to come apart, and if they do, it may result in breaking the parts. This type of fit is more challenging to create with less accurate 3D printers, but can still be achieved with some trial and error when testing the limits of your available resources.

A **transition fit** can be described as a snug fit between parts. You can press these parts together, as well as pull them apart, but there is little or no movement between them without a large amount of force. There is a low tolerance for this type of fit, but it's still easily achievable with most types of 3D printers and materials.

Lastly, a **clearance fit** has the biggest tolerance and the loosest fit. You will need to choose this if you want to create parts that can easily slide, rotate, or move when assembled. You could think of creating this type of fit as adding some "wiggle room" to your measurements to increase the play between parts within the design.

Modeling different fits in Tinkercad

Let's move into creating these different types of fits by changing the tolerance in our designs. We will begin by creating a socket and peg shape.

Creating a socket shape

To start, let's create a socket shape via the following steps:

1. First, add a **Cylinder** shape and dimension it to be 20 x 20 x 20 mm as shown in *Figure 12.6*:

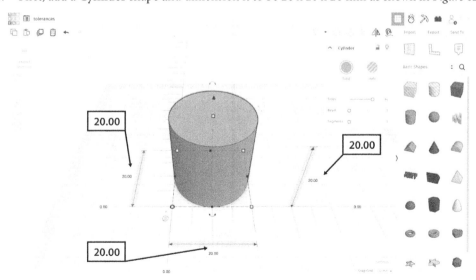

Figure 12.6: Starting with a cylinder in our design

2. Next, add a **Hole Cylinder** that is 15 mm long x 15 mm wide x 20 mm tall and raised 2 mm off the workplane as shown in *Figure 12.7*:

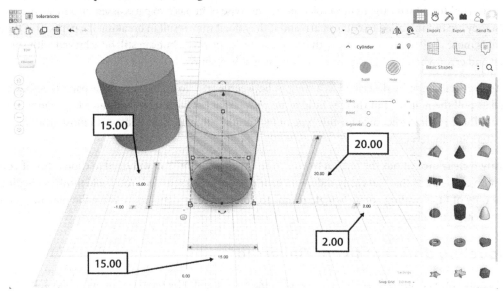

Figure 12.7: Adding a hole cylinder to our design

3. We can then use the **Align** tool to center the **Hole Cylinder** and **Solid Cylinder**, then **Group** the two shapes together as shown in *Figure 12.8*:

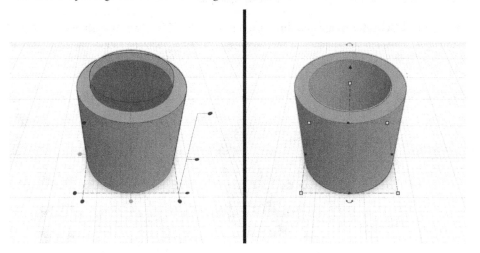

Figure 12.8: Aligning and grouping our shapes to create a socket shape

Next, we are going to make a peg shape that can fit into this socket shape.

Making a peg shape

We will create a peg shape to fit into the socket shape by using the different types of fits discussed earlier through the following steps:

1. Start by adding a **Cylinder** that is 20 x 20 x 3 mm as shown in *Figure 12.9*:

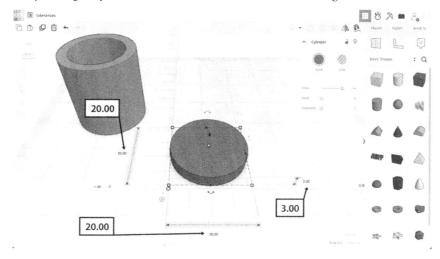

Figure 12.9: Starting with a cylinder in our design

2. Next, add a second **Cylinder** that is 15 x 15 x 20 mm and aligned to be centered in the first one, as shown in *Figure 12.10*:

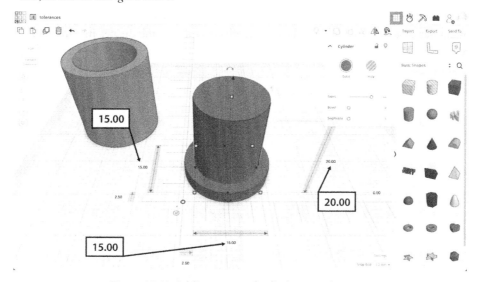

Figure 12.10: Adding a second cylinder to make a peg

3. We can then add a **Text** shape as a hole to the top of this cylinder peg to label its diameter as 15 mm, then **Group** all of the shapes together, as shown in *Figure 12.11*:

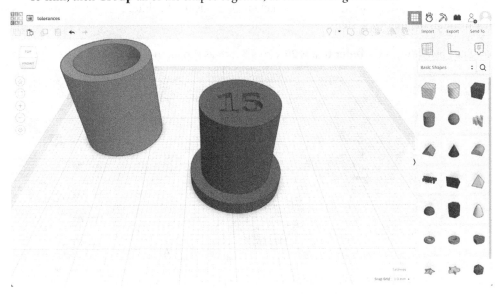

Figure 12.11: Creating a text feature as a label

Now we will 3D print and test different fits using the socket and peg shapes we just created.

Testing different Fits

Since the peg shape made in *Figure 12.11* is the same size as the hole made in *Figure 12.8*, this would be considered a press fit. No tolerances have been added, and pressing the peg into the socket would require quite a bit of force, as shown in *Figure 12.12*:

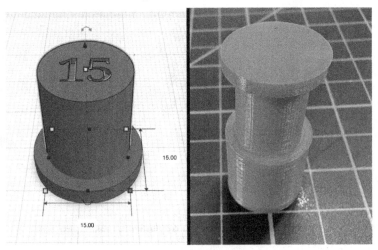

Figure 12.12: Testing the assembly of a press fit

The socket shape and peg shapes shown in *Figure 12.12* were printed on a FFF printer using PLA filament. As seen in this image, the peg struggles to fit into the socket shape. It's possible that this part would be able to fit tightly with a more accurate printer, but to make this model successful with this printer, I would need to adjust for the tolerances in my design as shown in *Figure 12.13*:

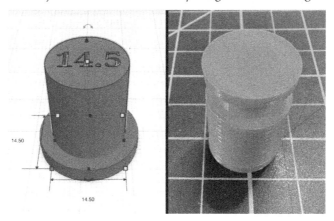

Figure 12.13: Adjusting and testing a transition fit

As shown on the left in *Figure 12.13*, I have adjusted the dimensions of the peg model to be 14.5 x 14.5 mm, making it 0.5 mm smaller than the socket to create a transition fit. This is how tolerances, such as the ones discussed in *Figure 12.3*, can be incorporated into our designs to create more wiggle room between our parts. I also adjusted the text shape to show this change in the model when it is printed as well.

As seen on the right in *Figure 12.13*, the peg now fits snugly into the socket with a bit of force and can also be removed without breaking the parts. Rotation is possible, but it's not quite the easy movement we would expect from a clearance fit, as shown in *Figure 12.14*:

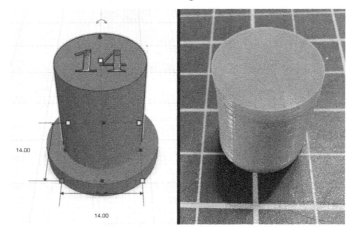

Figure 12.14: Adjusting and testing a clearance fit

As shown on the left in *Figure 12.14*, I've dimensioned the peg to be 14 x 14 mm, which will increase the gap between the peg and socket to create a looser fit. I've adjusted the text shape again to show this change in the model.

As seen on the right in *Figure 12.14*, the peg now fits easily into the hole to create a looser clearance fit, allowing for easy assembly and movement between the parts.

In practice, it is important to first determine the size that you want your parts to be and dimension them accordingly in your Tinkercad models. You must then consider the tolerances for the type of printer that you are using, as well as the dimensional accuracy for the material you are using, and compensate for these deviations in your designs. For more accurate printers, you may only need to adjust your model dimensions by 0.5 mm or less. However, when you're looking to create parts that can easily fit together through transition or clearance fits, the adjustments to your dimensions may need to be increased to compensate for the type of fit that you are looking to create.

It may take some trial and error to build proficiency in this. Fortunately, 3D printing is a powerful prototyping tool, and developing multiple iterations of your designs as you make these adjustments is an excellent way to learn. I've also created a model to test the tolerances needed for different fits by expanding upon the simple socket and peg designs created in this chapter, as shown in *Figure 12.15*:

Figure 12.15: A tolerance test model

There are 16 holes across the base of the model shown in *Figure 12.15* that increase in size relative to the peg from 0 to 1.5 mm. Whenever I find myself working with a new 3D printer or type of material, I print this model to get a sense of the tolerances needed for the type of fit that I am looking to create between my parts. As discussed earlier, there are other factors that may impact the accuracy of your models in addition to the printer and material, but I find something like this to be a good starting point to help increase the effectiveness of my designs.

Applying tolerances in a real-world setting

We should continue to consider tolerances, even if our multi-part designs do not involve multiple 3D-printed parts. Let's say that we want to make a 3D-printed part that interacts with an existing product, such as a pencil or marker, by creating a 3D-printed organizer. We could work to create a model that supports these products based on the skills learned in this chapter through the following steps:

1. We will first want to measure the product that we will be working with, as shown in *Figure 12.16*. As seen in this image, we can use digital calipers to determine that the diameter of my part is 11.5 mm.

Figure 12.16: Measuring a marker to determine its diameter

2. We can create a **Hole Cylinder** that can fit our part by adding to the dimensions to create a clearance fit (12.5 mm), as shown in *Figure 12.17*:

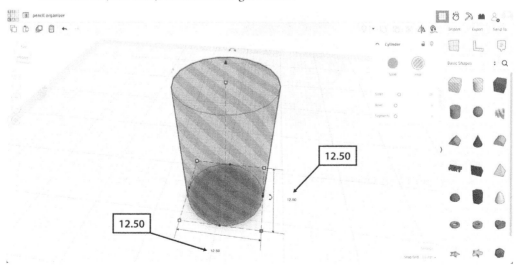

Figure 12.17: Creating a hole shape dimensioned to be a clearance fit

3. We can then raise this hole shape so it is 2 mm off the workplane, as shown in *Figure 12.18*:

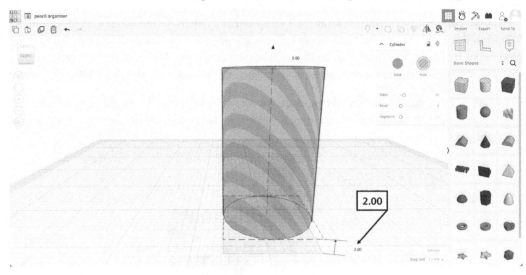

Figure 12.18: Lifting the hole shape off of the workplane

4. Next, we can align this hole to be centered within a **Solid** shape, as seen in *Figure 12.19*:

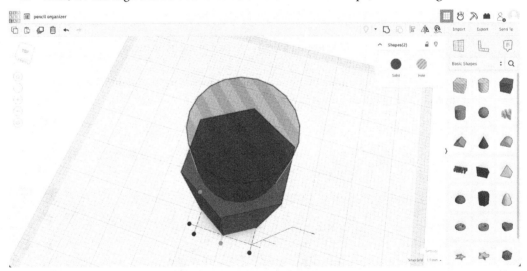

Figure 12.19: Adding a solid shape and aligning it to the existing hole shape

Adding tolerances to our designs | 197

5. We can then **Group** these two shapes to create a solid part, as shown in *Figure 12.20*:

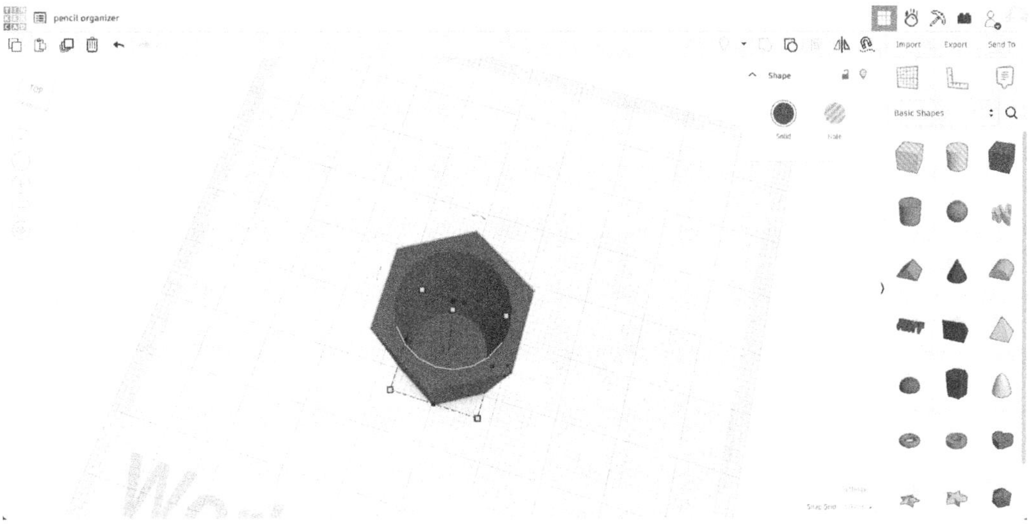

Figure 12.20: Grouping the two shapes to create a solid part

6. This could be our prototype, as it should allow for the part measured earlier to fit within the model. However, let's say that we would like to expand this prototype by copying and pasting more of these shapes and arranging them together, as shown in *Figure 12.21*:

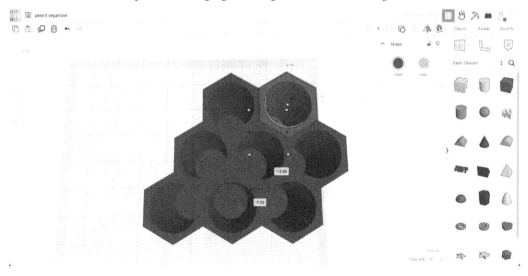

Figure 12.21: Copying and pasting the solid shape to make multiples in the design

7. To add to the aesthetics of this model, we can then change the heights of each copied part, ranging from 55 to 35 mm, to arrange them in a more aesthetically pleasing way, as seen in *Figure 12.22*:

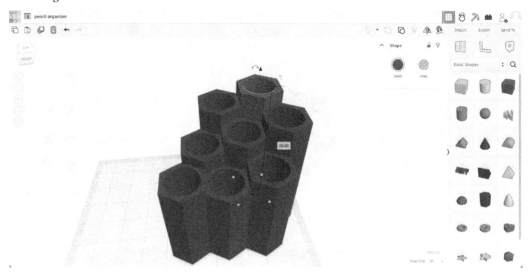

Figure 12.22: Varying the heights of our parts to change the overall appearance

8. After 3D printing this model, we can test the fit and performance, as seen in *Figure 12.23*:

Figure 12.23: Testing the performance of the 3D printed organizer

This marker holder demonstrates the importance of tolerances when creating a simple 3D-printed project, as failure to add tolerances to this design would have prevented the markers from being able to fit into the holder. As our designs become more complex, we must continue to consider tolerances whenever we create different fits between our parts. And when all of the parts in a project are 3D printed, we must take greater care in determining the tolerances, as each part's final dimensions will vary. We'll dive deeper into these concepts later on, such as when creating a multi-part box in *Chapter 16*.

While we can discuss best practices and common strategies for implementing tolerances in our design, it is important to remember that there are many circumstances wherein your dimensions and tolerances may need to change. It's important to maintain an awareness of tolerances as you strive to make multi-part designs, and also to work to test and adjust your measurements as needed in order to create more effective 3D printed products and prototype solutions.

Summary

We learned that tolerances are an important factor in all forms of manufacturing and one that we must consider as we strive to manufacture our complex models effectively using 3D printing technology. While there isn't a single correct approach to accommodating tolerances in designs, we learned that there are general strategies and common practices that can be employed to find success.

This chapter mentioned that the first step is to test and determine the tolerances that can be expected from the resources we have available to us. We learned that the type of printer, the type of material, and even the printing environment and CAM software settings can all change the achievable tolerances and accuracy. Creating test models and prototyping designs is an effective approach to determine these factors, as mentioned in this chapter.

We also learned that we can then incorporate these tolerances into our Tinkercad designs by accommodating them within our measurements. By adding some wiggle room to our dimensions, we can effectively design models that have parts that fit together tightly, or loosely, based on our desired fit.

As we strive to expand our skills and abilities in creating complex models with Tinkercad that are 3D printed using a wide range of techniques, further building upon our skills in taking measurements, adding dimensions, and testing our models is key. Throughout this process, we should continue to apply other best practices, such as the ones discussed previously in *Chapter 11*, as well as avoid the common mistakes that we will be discussing in the next chapter. When you're ready, move on to the next chapter to learn more about things that we should avoid doing in our designs.

13
Design Mistakes to Avoid

Over the past few chapters, we have been looking at strategies and techniques for effectively designing models that can be 3D printed using Tinkercad. In this chapter, we will be looking at some of the most common design mistakes that are easy to make as we strive to create our models, and how to avoid making them.

Throughout this chapter, examples and techniques will be broken down into the following topics:

- Watching the workplane
- Identifying thin lines and walls
- Connecting our parts

This chapter will allow us to not only continue to expand our skills and abilities in learning how to use Tinkercad for 3D printing effectively but also gain greater confidence and ability in checking our designs for possible mistakes, as well as striving to avoid making them altogether.

Technical requirements

As we learn about common mistakes that can be made in our designs, we will continue working with Tinkercad as we have been throughout this book. Many of the strategies we will be discussing are applicable to a wide range of 3D printing techniques and material choices, so not knowing which resources you may have available just yet is OK and will not prevent you from finding success with the topics we will be covering in this chapter.

We will also be looking at how some of these mistakes can be identified and corrected in CAM software for 3D printing later in this chapter. There are many different CAM programs, also known as **slicers**, available for 3D printing as initially introduced in *Chapter 10*. Choosing a CAM program will depend on the model and type of 3D printer you are using, as we will discuss later in *Chapter 14*.

You can also find an editable version of the Tinkercad model referenced in this chapter at this link: https://www.tinkercad.com/things/5dW1pyOip2i-mistakes-model-from-chapter-13

Watching the workplane

The first design mistake we are going look at is one that could be catastrophic if not addressed, and one that I find easy to make when designing 3D models in Tinkercad. This mistake has to do with ensuring our models are touching the workplane.

As discussed in *Chapter 11*, having a larger amount of surface area touching the workplane increases the likelihood of success for our first layer, and the subsequent model built upon it. Failing to have our models touch the workplane, like the one shown in *Figure 13.1*, would not print successfully.

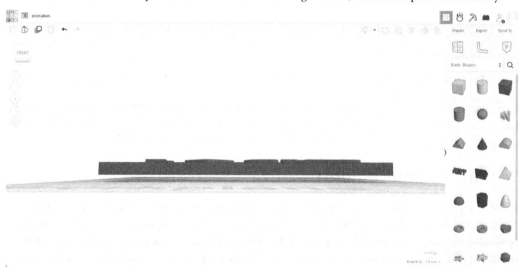

Figure 13.1: Looking at a model that is floating above the workplane

When looking at the model shown in *Figure 13.1* from the front, it is obvious that this model is floating above the workplane. If we were to try to print this, the first few layers would actually print nothing, before then printing the model suspended in mid-air, which would be a mess.

But it is easy to miss a mistake like this if we do not utilize our navigation tools in Tinkercad. For example, *Figure 13.2* shows the same model from a top-right perspective view:

Figure 13.2: Looking at a floating model from a different perspective

As discussed in *Chapter 3*, it is important to **orbit** around our model to inspect it from multiple perspectives before exporting it for 3D printing. Failing to do this might cause us to miss simple mistakes like these, which may have a catastrophic effect during production. Once we catch a floating model, we want to use our transformation tools to lower it so it is touching the workplane, or we can also press *D* on our keyboard to automatically drop our model to the workplane as well.

Something that is also easy to do is to accidentally rotate our models so they are only partially touching the workplane, as shown in *Figure 13.3*:

Figure 13.3: Looking at a model that has been rotated slightly

Initially, it may not be clear that the model shown in *Figure 13.3* is off the workplane when looking at the image to the left. But when we zoom in, as shown on the right, it is more clear that there is a slight gap between the model and the workplane on one side due to a slight rotation.

Unlike fully floating models, like the ones shown in *Figures 13.1* and *13.2*, this mistake can be more difficult to catch, and pressing *D* won't fix it because the model is already touching the workplane, just not fully. To correct this, you could attempt to rotate the model back so it is flat manually using the rotation tool by degree, or press **Undo** if you happen to catch it in the act. You can also check the model in your CAM software, as seen in *Figure 13.4*:

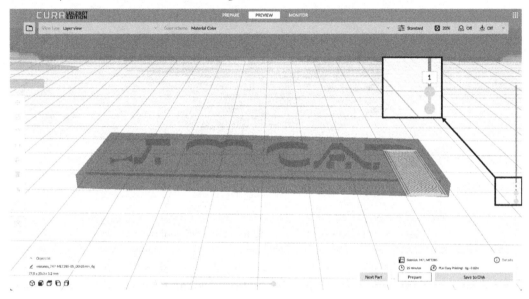

Figure 13.4: Inspecting the first layer of a model in Cura LulzBot Edition

All slicer programs, such as *Cura LulzBot Edition* shown in *Figure 13.4*, allow you to simulate and preview how your model will actually be printed. In this preview, we can inspect the first layer to ensure it looks as it should. When looking at *Figure 13.4*, we can see that there is only a small part of our model touching the build plate in layer **1**, but when we move up a few layers, we see something different, as shown in *Figure 13.5*:

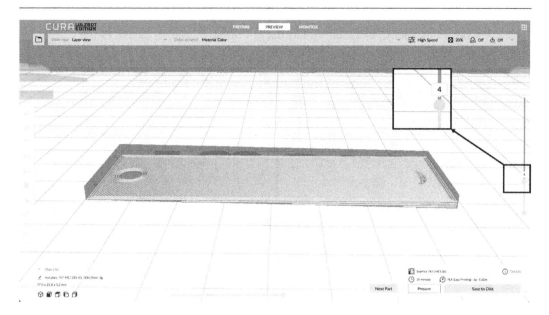

Figure 13.5: Inspecting the fourth layer of a model in Cura LulzBot Edition

When looking at layer **4** in *Figure 13.5*, we can see that a much larger area is now touching the workplane than what we saw in our first layer shown in *Figure 13.4*. This is a clear indication that our model must have been rotated slightly, and we didn't catch it in the design stages. Depending on the CAM program you are using, some additional tools to address mistakes such as floating or rotated models may be available without needing to return to the CAD software to fix the model.

As discussed in *Chapter 11*, we could add support material below the model to support the area suspended above the workplane, which could then be removed after printing, though this wouldn't be ideal. Adding support for an issue like this may reduce our overall print quality and add steps to the postprocessing stages of production.

Some CAM programs, such as Cura, offer a **Lay flat** or **Bottom** button, as seen in *Figure 13.6*:

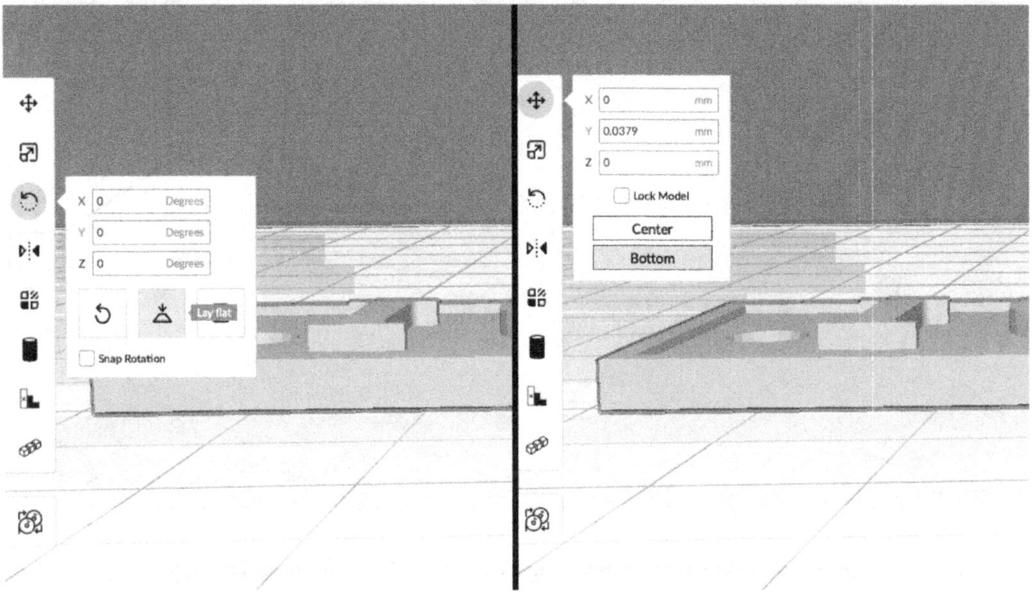

Figure 13.6: Options to rotate and move a model in Cura LulzBot Edition

As shown in *Figure 13.6*, this version of the Cura 3D printing CAM program has tools to automatically rotate models so that they are flat and touch the workplane. This is a nice feature to have so we can correct this mistake easily before printing, but it is still best to check your model and correct mistakes like these in the CAD stages before exporting them.

There are many other things we can do in CAM programs to support our models for 3D printing, much of which will be discussed further in the next chapter. In the meantime, let's move on to another common mistake, creating lines and walls that are too thin to print.

Identifying thin lines and walls

In *Chapters 10* and *12*, we discussed the quality and detail we could expect from common types of 3D printing techniques, such as FFF, FDM, and SLA. As discussed, the resolution achievable by a 3D printer is typically defined as the **layer height**, and the smaller the layer, the smoother the print and the more detailed it will be.

But what about resolution in the other dimensions? Layer height is a factor in the vertical dimension, or Z, and we should also be concerned with how small our printers can create **lines** and **walls** in our design. Let's say we want to create a ring, for example, as shown in *Figure 13.7*:

Identifying thin lines and walls

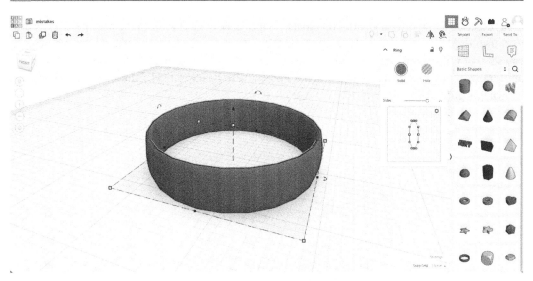

Figure 13.7: The Ring shape in Tinkercad

The **Ring** shape is shown in *Figure 13.7*, and initially, we might not question whether or not this can be 3D printed. But something we need to check in our design before attempting to 3D print our models is the wall thickness, which is highlighted in *Figure 13.8*:

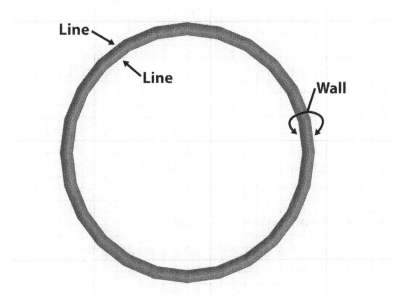

Figure 13.8: Identifying lines and walls of the ring shape

Walls describe the structural features that make up the shape of our model in the *X* and *Y* dimensions, as shown in *Figure 13.8*. When looking at the **Ring** shape from a top view like the one shown, we can see that walls are made up of multiple lines with a space in between. Wall thickness changes the appearance and performance of our designs, and can also be impacted by the shapes we choose, as shown in *Figure 13.9*:

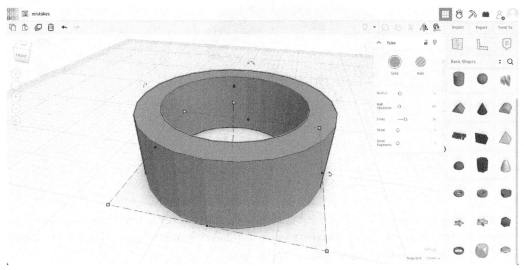

Figure 13.9: Analyzing the walls of a Tube shape

The walls of the **Tube** shape shown in *Figure 13.9* are thicker by default than the ones in the ring shape shown in *Figure 13.7*. And unlike the ring, we have direct control over the thickness of the walls for the **Tube** shape using the shape parameters shown.

To be able to manufacture thinner walls and lines, we would need to have a higher-resolution 3D printer, just like if we wanted to create thinner layers. But what is too thin? Well, like layer height, this may vary based on the type of 3D printer you are using. Let's consider how our walls should be adjusted if we are designing for an extrusion-type or resin-type printer, as well as if we are looking to make walls that are stronger or higher performing too.

Working with extrusion-type printers

For extrusion-type printers, such as FFF and FDM, the minimum wall and line thickness is usually determined by the diameter of the nozzle, as shown in *Figure 13.10*:

Figure 13.10: Looking at a nozzle of an FFF-type 3D printer

Figure 13.10 is a close-up image of the nozzle on the extruder for an FFF-style 3D printer. We can see the opening of the nozzle, which is a very small hole where the filament is fed through during the printing process. The diameter of this opening determines not only how quickly filament can be extruded but also how thin lines and walls can be created. The nozzle shown in *Figure 13.10* is 0.5 mm in diameter, which can therefore create a line of 0.5 mm in width. If I were using a 3D printer with a bigger nozzle, say 1.2 mm, the minimum line width would then also be much bigger, as shown in *Figure 13.11*:

Figure 13.11: Comparing minimum line widths of different-diameter nozzles

Figure 13.11 compares the minimum line widths of a 0.5 mm nozzle and a 1.2 mm nozzle, and it's clear that there is a significant difference in line thickness based on the nozzle chosen. Thicker isn't worse necessarily; it depends on the type of model that you are looking to create. The thicker lines made by the 1.2 mm nozzle shown in *Figure 13.11* may not be able to achieve the detail that the thinner lines could, but they did print faster and are also far stronger than the thinner lines too. The consistency of the line and wall thickness may also be impacted by the motors and movement system that allows the extruder and bed to move back and forth as well, but these systems typically have a bigger impact on the overall quality of your models rather than the individual lines and walls created.

But what about SLA or other resin-type 3D printers?

Working with resin-type 3D printers

As we learned in *Chapter 10*, there are no nozzles in this printing technique, so what determines the minimum line and wall thickness achievable? This is typically determined by a combination of the resolution of the light source used, the resin chosen, as well as the mechanical constraints of the motors and movement systems used. Typically, SLA 3D printers can create a minimum wall thickness of 0.2 mm.

However, it is important to note for both extrusion-type and resin-type 3D printing techniques, just because the minimum wall thickness may be printable, it doesn't mean that our model will perform as intended at these minimums. Minimum wall thicknesses are usually very delicate and prone to breaking and warping. As discussed in *Chapter 12*, we often want to add a **tolerance** to our minimums, which is typically recommended to be +0.5 mm. So, for extrusion-type printing methods such as FFF and FDM, a recommended minimum wall thickness would be approximately 1.0 mm, and for resin-type printers such as SLA, 0.7 mm would be a recommended minimum.

Knowing what tolerances we can expect based on the type of printer we are using will help us design more successful models, but we should also consider how the walls are created as this too may impact wall performance and tolerances.

Adjusting the performance of our walls

Something else to consider that impacts the performance of our walls is whether or not the wall is supported, as shown in *Figure 13.12*:

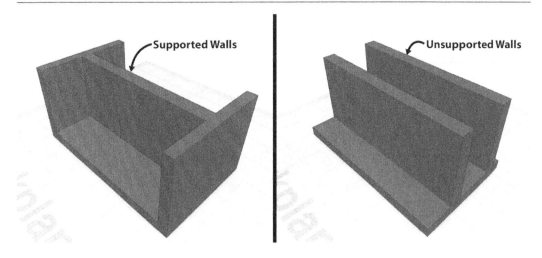

Figure 13.12: Comparing supported and unsupported walls

A supported wall is a wall that is connected to another wall, as shown in the model to the left of *Figure 13.12*. Unsupported walls, as shown to the right, are walls that are freestanding. Having walls that are supported or unsupported does not change the minimum of what your 3D printer can achieve, but it will change how your model performs at these minimums. A supported wall is a much stronger wall, and therefore would require lower tolerances to perform without warping or breaking when compared to an unsupported wall.

So what if I did want to make the thin ring shown in *Figure 13.7*, or even the tube shown in *Figure 13.9*? I would need to determine whether the walls of the model are thinner than our recommended thicknesses based on the type of printer I was using. And if they were, lines and walls that are too thin are typically ignored altogether by the CAM program during the slicing process and ones that are at the bare minimum are likely to break during or after the printing process. As we design models, we should always measure our wall thicknesses to ensure they are at or above the minimum of what our 3D printer can achieve, and design our models to support the walls where possible to ensure greater success.

But checking to ensure our walls are connected covers just one part of our designs. We should also inspect the other parts of our models too.

Connecting our parts

The last mistake we are going to look at is a common one to make when combining multiple shapes to make our models, as shown in *Figure 13.13*:

Figure 13.13: Analyzing a Tinkercad model made from multiple shapes

At first glance, there may not appear to be anything wrong with the model shown in *Figure 13.13*. It is made from multiple shapes, including boxes, text, and a cylinder as a hole shape. Before downloading this model to be 3D printed, it is important to look around our model to make sure all the shapes used are connected, as shown in *Figure 13.14*:

Figure 13.14: Viewing a model from a front perspective

Figure 13.14 is now looking at the front of the same model, and in this view, it is clear that the text is actually hovering above the design rather than touching it. If we were to try to print this, and if support material were not enabled, this text feature would be floating and would fail to print correctly. If we continue to orbit around our model, we can inspect it for other issues, as shown in *Figure 13.15*:

Figure 13.15: Viewing a model from a top perspective

Now, looking at the top of the same model in *Figure 13.15*, we can see there is actually a very small gap between the outer and inner boxes. This was hard to see when zoomed out in our first perspective view, but it has become clear when looking more closely at this design. If we were to try to print the design without connecting these two boxes, it would probably print OK, but then easily break apart when in use.

Mistakes such as these are easy to make because of the nature of how models are created in Tinkercad. We must ensure that the shapes added are touching without gaps because these gaps will remain during the printing process, even if our shapes are grouped together. After connecting our shapes, we want to make sure that they are grouped together as missing this step could lead to another mistake during the exporting process, as shown in *Figure 13.16*:

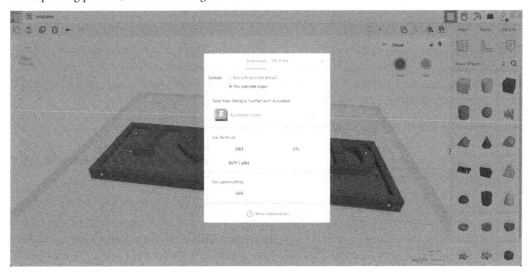

Figure 13.16: Viewing options for exporting a model in Tinkercad

After pressing **Export** in the top-right corner of the screen, we are shown the **Download** window, as seen in *Figure 13.16*. This is where our model can be exported for 3D printing after completing the model. Notice how only **The selected shape.** is highlighted in the window shown. If I were to export like this, I wouldn't download the entire model but instead just the single or couple of shapes I had selected before pressing **Export**.

This is a convenient feature if you have multiple parts or models in your design and you want to download them individually. But if you are creating a single model and forget to group your shapes together before exporting them, you may download them as individual shapes rather than the completed model as intended. By grouping our shapes together first, we can ensure the entire model is downloaded during the export process without needing to choose **Everything in the design**. Choosing this option would download all shapes regardless of whether they are grouped or not, but it would also download any additional shapes or models that we may not have wanted to include.

Now that we have looked at common mistakes that are easy to make during the design and exporting process, it's time to look at what to do when exporting our designs and how we could get them onto our 3D printers in the next chapter.

Summary

Throughout this book, we have been learning complex skills and techniques to not only design unique models in the Tinkercad program but also design models that could be 3D printed successfully and effectively. As with anything, making mistakes is a natural part of the design process and one that you should not be afraid of, as this is part of learning and improving your skills. But there are some mistakes that are easy to make, and an important step to learning is how to identify and check for these mistakes as they may greatly impact the success of your designs.

Remember to always ensure your models are touching the workplane, and not accidentally rotated so that they are only partially touching the workplane. We can inspect our models from multiple perspectives in Tinkercad to check this, and also check what our first few layers look like as we prepare our designs in CAM later on as well.

When designing our shapes, we should consider the type of printer we are using to ensure that the lines and walls within our design are not too thin to be manufactured successfully. When possible, we should work to create supported walls and add tolerances to the wall thickness so that they are not only printable but also perform effectively too.

Before exporting our designs, we should always orbit around our model to view it from multiple perspectives. We want to not only check the workplane but also check to ensure all shapes in our designs are connected and touching without any gaps between them. Before exporting, we should always group our shapes if they are intended to be connected and then export them as a single model as well.

Now that we've identified these common mistakes, we can work to check for them and avoid them as we use Tinkercad to create models for 3D printing more effectively. When you're ready, turn the page to put these skills to the test as we proceed with exporting our models to bring them into CAM software for production in the next chapter.

14
Exporting and Sharing Tinkercad Designs for Manufacturing

In this chapter, we will look at how we can take our Tinkercad designs off the screen and into the physical world by exporting and sharing our designs for manufacturing. As discussed in the past few chapters, there are many different types of 3D printers that we could use to manufacture our designs, as well as strategies and factors that we must consider during manufacturing to find greater success. We will work to put all of these concepts together as we break this chapter down into the following topics:

- Exporting our designs
- Choosing and using CAM software
- 3D printing Tinkercad designs with Autodesk Fusion
- Finding 3D printing services

Through these topics, we will look at key strategies and resources as you choose the methods that work best for you and your unique workspace. We will also take a look at the services that are available to help you manufacture your designs, even if you don't personally own a 3D printer.

By the end of this chapter, you will be able to create connections between the key topics and skills covered earlier in this book as we move into the final stages of the design and production process.

Technical requirements

To prepare a design to be manufactured using 3D printing technology, a CAM program will need to be used. CAM programs vary from printer to printer, so it is important to identify which CAM program is needed for the resources you have available. Unlike Tinkercad, many CAM programs are not web-based and would need to be installed on a Windows, Mac, or Linux computer.

There are generic CAM programs that work with a wide range of 3D printers, such as **Slic3r** (`https://slic3r.org`), **Cura** (`https://ultimaker.com/software/ultimaker-cura/`), and **Autodesk Fusion** (`https://www.autodesk.com/products/fusion-360`). If you are constrained to using a tablet or Chromebook device, Fusion and **Kiri:Moto** (`https://grid.space/kiri/`) both offer a browser-based interface like Tinkercad. But if you don't have a 3D printer of your own, we will also be looking at 3D printing services available through Tinkercad later in this chapter as well!

Lastly, you can access and editable version of the sample Tinkercad design shown throughout this chapter at this link: `https://www.tinkercad.com/things/fcJsp1JBWzb-sample-part-from-chapter-14`

Exporting our designs

The first step in manufacturing your Tinkercad model is to export it into a generic file that can be read by our CAM programs, also known as **slicers**, as originally introduced in *Chapter 10*.

To do this, start by creating or opening a 3D design in Tinkercad and find the **Export** button in the top-right corner of the design window as shown in *Figure 14.1*:

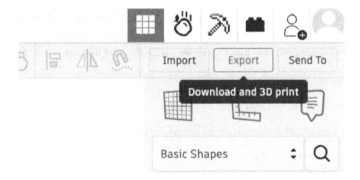

Figure 14.1: Find the export button in Tinkercad's 3D Design space

As shown in *Figure 14.1*, the **Export** button is located at the top of the toolbar on the right side of our design window. We can press **Export** even if we have not selected any shapes in our design, but you should select a shape first if you only want to export a part of your design rather than everything. These options are also shown in the export window, as seen in *Figure 14.2*:

Exporting our designs 217

Figure 14.2: The options for exporting a design in Tinkercad

If we wanted to download our models to 3D print them ourselves, we would have a few different file types that we can choose from, as seen in *Figure 14.2*. Choosing the right file format is an important step depending on what type of CAM program you will be using, as well as how you might plan to share your designs. The differences between these file formats are as follows:

- **Stereolithography** (.STL): This is the most common and generic type of 3D file used for 3D printing. An .STL file stores our models as a set of vertices joined by edges to make triangular faces. Model color is not retained in an .STL file.

- **Wavefront OBJect** (.OBJ): A more complex file type when compared to .STL, .OBJ files are still widely used as a generic file type for 3D printing. An .OBJ file retains models as vertices like .STL files do, but also support polygons, which typically offer a more detailed representation of the original model. Model color is not retained in an .OBJ file.

- **Graphics Library Transmission Format** (.GLB): A .GLB file is less common for 3D printing, but is commonly used for transferring and using 3D files in animations and games, as well as for web use. Downloading our models as a .GLB file will retain the colors we used and allow us to bring our models into different programs for display and viewing, including *PowerPoint*!

Once you have identified which file type will work best for your needs, click on that file type to start the download process. Complex files may take a few minutes to download to your computer, and we can see that a file is being processed by the message shown in *Figure 14.3*:

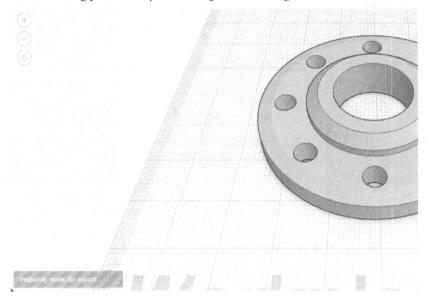

Figure 14.3: The status message for downloading files

As a file is being downloaded, it is important to keep your Tinkercad design window open and avoid making any changes until the download has finished and the message shown in *Figure 14.3* goes away. Once the download has been completed, you will notice that the name of the exported file matches the name of your Tinkercad design.

Depending on your device, you may not be able to open or preview a design once it has been exported in the file formats listed. To view an exported design, you may need to open it in a 3D file viewer, in a CAD program such as Tinkercad, or in the CAM program you will be using to 3D print, which we will discuss in the next section.

Choosing and using CAM software

When you buy a 3D printer, it is very likely that the manufacturer will provide you with a slicer program to use with your specific model printer. Some printers may only work with the slicer program supported by the manufacturer.

However, many printers are compatible with a generic slicer program such as Slicr, Cura, or Kiri:Moto. In fact, you might find that your printer comes with a modified version of these generic slicer programs as they are often **open source** applications. An example of this would be **Cura LulzBot Edition**, which is a modified version of Cura adjusted specifically for the features included with LulzBot 3D printers.

However, all of these programs, even the brand-specific ones such as Cura LulzBot Edition or PrusaSlicer by Prusa Research, can often still work with printers made by other manufacturers. Thanks to this flexibility, you have the freedom to browse and choose a slicer program that works well for your needs and is compatible with both your printer and your device.

Throughout this section, I will be demonstrating how to prepare a design exported from Tinkercad using the commonly used Cura slicer by Ultimaker. You can download and use Cura for free (refer to the *Technical requirements* section), and while the following steps are completed in this version of Cura, the steps for preparing a 3D model for 3D printing are very similar in most slicer programs.

Setting up Cura

After installing and launching Cura, we will be shown a window asking us to select our 3D printer from a list of printers made by Ultimaker, or to add another printer, as shown in *Figure 14.4*:

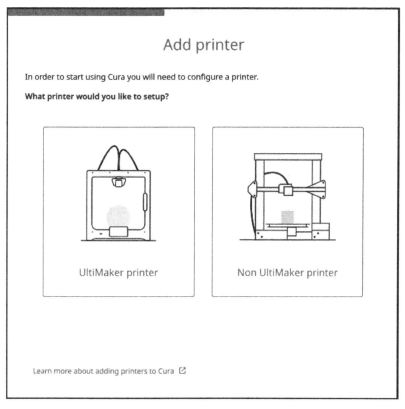

Figure 14.4: The window to add a printer in Cura by Ultimaker

After choosing or adding a printer, we can find the open file button, which looks like a file folder to open our exported model from Tinkercad in Cura, as shown in *Figure 14.5*:

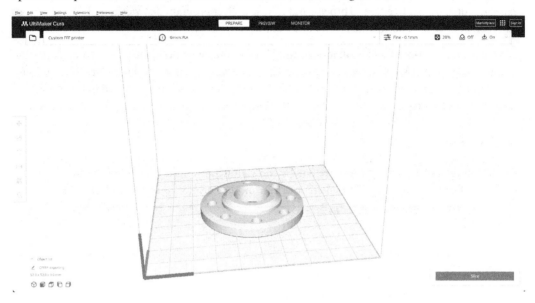

Figure 14.5: Loading a 3D model file exported from Tinkercad

After loading a model, you will not be able to change the features or shapes within your design as the files that we exported are not editable in CAM. To make any design adjustments, you would need to return to the original Tinkercad design, make any necessary changes, and then export your model again. CAM programs typically have some tools to adjust the scale or position of our models though, as shown in *Figure 14.6*:

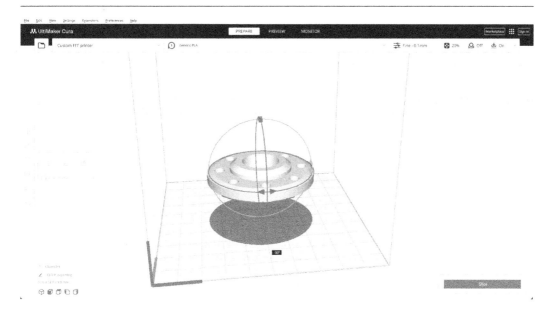

Figure 14.6: Adjusting the orientation of a model in Cura

If needed, we can move or rotate an imported model on the build plate in our slicer, as shown in *Figure 14.6*. Some slicers also allow you to split a part up into multiple files, as well as import multiple models to print at the same time.

Once we've adjusted our model position as needed, it is time we prepare our design for 3D printing. This is where some of the important settings we discussed in *Chapters 10* and *11* come into play to ensure that we find success.

Preparing a design in Cura

The first step to preparing our designs for manufacturing is choosing the material that we will be using from the material dropdown shown in *Figure 14.7*:

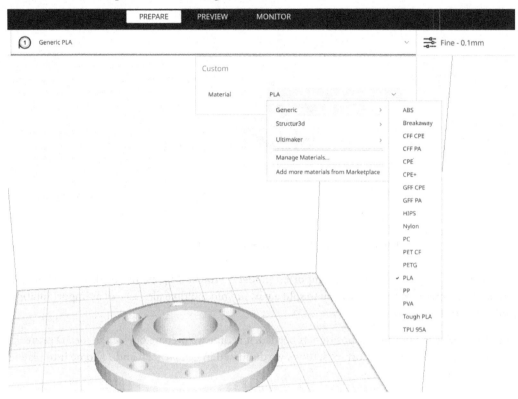

Figure 14.7: Adjusting the material in Cura

All slicers will come preloaded with a list of filaments or materials that can be used to prepare your models for printing. In general, most materials and brands have universal settings, meaning one roll of PLA will print like another roll of PLA from a different manufacturer. However, it is always best to choose or tune the settings to match the specific material you are using whenever possible.

Adjusting the print settings

We can adjust the print settings in the settings dropdown, as shown in *Figure 14.8*:

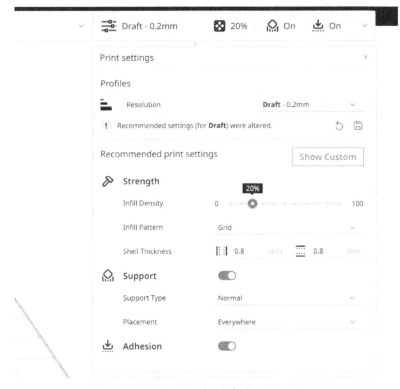

Figure 14.8: Adjusting the default print settings

As we prepare our models for 3D printing in the settings window shown in *Figure 14.8*, some things to consider are as follows:

- **Resolution**: This selects the layer height and detail of our 3D prints. A larger layer height will print faster but also produce a lower-quality model. The options for layer height will vary based on the 3D printer you have chosen and will also vary from printer to printer.

- **Infill density**: This lets you choose how dense, or hollow, a model is. A higher density will use more material and make a more solid part, while a lower density will cut down on print time and material usage, but also make a hollower and potentially weaker part. The default recommended density is usually around 20%.

- **Infill pattern**: You can change the shape of the infill pattern that fills your model. Choosing different infill patterns will change the print speed, material usage, and method in which a part performs. The default option is typically lines or grid, but you can choose a different infill pattern to support the unique shapes you might have created in Tinkercad.

- **Shell thickness**: This determines the minimum thickness for your walls, floor, and roof, as discussed in *Chapter 13*.

- **Support**: This is where we can choose to enable support material as discussed in *Chapter 11*. If your model has overhangs or complex structures, enabling support may be required for printing success.

- **Adhesion**: This sets supports specifically for better adhering our prints to the build plate, as discussed in *Chapter 11*. When disabled, a **skirt** will be printed as a perimeter around your design to show where the part will be and to ensure that the nozzle is extruding correctly. When enabled, a **brim** will attach this skirt to the model for increased adhesion.

You will find that there are default parameters for all of the settings in any slicer you are working with and that many of these defaults serve as a good starting point. They might not give you the fastest or most efficient print, but they should work well to get you going and help you find initial success. As you expand your abilities, you can dive deeper into hundreds of custom settings, as shown in *Figure 14.9*:

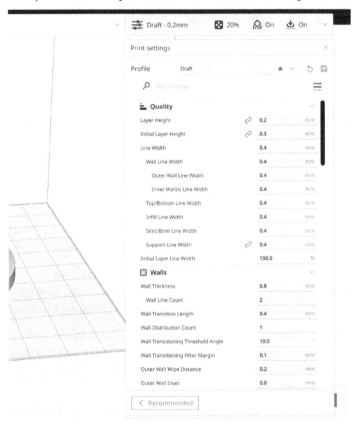

Figure 14.9: The custom settings window in Cura

There are many things that can be tweaked or adjusted in the custom settings, from print speed to how the walls are actually printed and much more. I recommend that you start small and make little adjustments as you experiment with finding optimal settings for your needs. After making some custom adjustments, you can save your settings as a default, as shown in *Figure 14.10*:

Figure 14.10: Saving a custom profile in Cura

Custom profiles, such as the one created in *Figure 14.10*, allow you to quickly return to custom settings you have created over the defaults. However, these profiles are unique to the printer and material you have selected, so you will need to re-adjust and continue to experiment each time you change the resources you are using.

Many slicers, such as Cura, also do a good job of warning you if one of your parameters may not work. Cura does this by highlighting the setting, as shown in *Figure 14.11*:

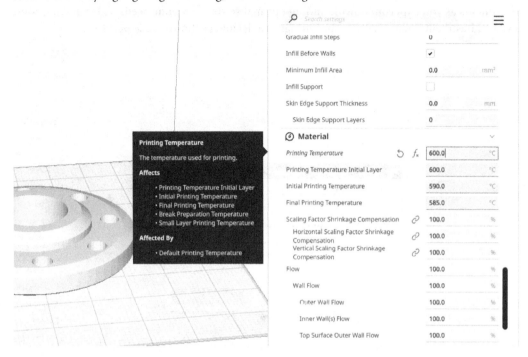

Figure 14.11: A warning for a parameter set in Cura

When something such as the temperature is set wrong, as shown in *Figure 14.11*, Cura will highlight this as a warning to allow us to see whether what we are adjusting may be wrong or an error before accidentally sending a file to our printers that won't work. However, it's important to note that these warnings are generic and may not appear every time we adjust something incorrectly. As such, you still need to use your best judgment when making tweaks to your profiles.

Applying the adjusted settings

Once we've adjusted our settings as desired, we can press the **Slice** button to apply these settings to our model as we prepare it for 3D printing, as seen in *Figure 14.12*:

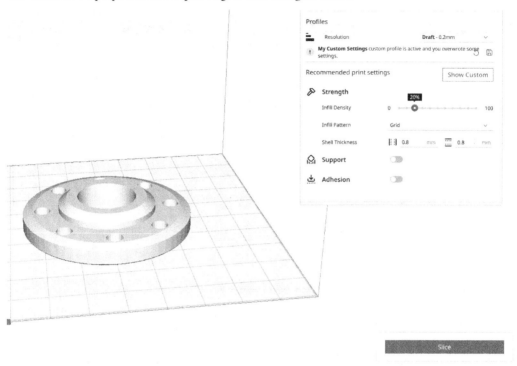

Figure 14.12: Pressing Slice to prepare our models for printing

Some slicer programs may slice your model for you automatically as you make adjustments, so pressing the button shown in *Figure 14.12* may not be needed. However, once the model has been sliced, we can press **Preview** to see how it will be printed, as well as how long it will take and how much material it will use, as shown in *Figure 14.13*:

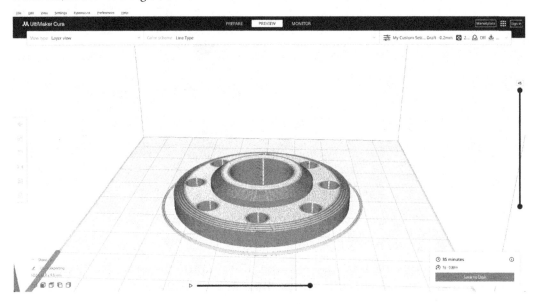

Figure 14.13: Previewing a sliced model in Cura

During this preview stage, it is important to inspect your model for any flaws, such as designs that are floating above the build plate or ones with walls that are too thin, as discussed in *Chapter 13*. We can also check to make sure that we have enough material loaded in our printers to complete this print, as an approximate amount of filament needed is shown in the preview, as seen in *Figure 14.13*.

In addition to the print time and material used listed in the details window shown in *Figure 14.13*, we can also send this prepared model to our 3D printer, which is the final step. If your printer is connected via your network and linked to your CAM program, you should see an option to send this file to your printer in this window. Alternatively, you may need to press **Save to Disk** to save the file to a flash drive, SD card, or other disk, as shown in *Figure 14.13*.

The type of file we created in our slicer is called a **Gcode** file. It is a coded file that instructs our 3D printers on how to make our models using the settings we've chosen, as discussed earlier in *Chapter 10*. You can preview the Gcode file created in a basic text editing program and even write your own custom Gcode if you are well-versed enough to do so! Once we have loaded this Gcode file onto our printers, we will be able to see our Tinkercad models come to life, as shown in *Figure 14.4*:

Figure 14.14: Printing a design made in Tinkercad and prepared in Cura on a LulzBot FFF Printer

An alternative to using a CAM program designed specifically for 3D printing, such as Cura, is to send our Tinkercad designs to a more versatile and advanced design program such as Autodesk Fusion.

3D printing Tinkercad designs with Autodesk Fusion

Autodesk Fusion is a unique program as it combines CAD and CAM into one application. You could use Fusion to design 3D models, as we have learned to do in Tinkercad, and you can also use Fusion to prepare models for 3D printing, like we did using Cura in the previous section.

A benefit of using Fusion is that it allows for a seamless transition from Tinkercad, as both programs are made by Autodesk. Fusion also offers advanced modeling tools, as well as tools for rendering and animation, which may allow you to elevate your design and modeling skills past what is possible in Tinkercad. And unlike CAM programs like Cura which are specific to 3D printing, Fusion's CAM features are compatible with a range of additive, subtractive, and formative manufacturing techniques as we discussed in *Chapter 10*.

However, in this section, we will specifically look at how we can use Fusion to manufacture our Tinkercad designs through 3D printing techniques.

Sending Tinkercad designs to Fusion

Before being able to send a design to Fusion, you need to create an Autodesk account, which is free for hobbyists, students, and teachers. You also need to either access Fusion via the web app or install the desktop version on a Windows or Mac device. Once Fusion has been installed and your account has been created, we can return to Tinkercad and once again open the export window as was shown earlier in *Figure 14.2*.

In the window shown in *Figure 14.2*, we previously downloaded our model to bring it into a slicer program such as Cura. Now, we will press the **Autodesk Fusion** button to transfer our Tinkercad designs to the Fusion program. Once we press this button, a new window will appear, as shown in *Figure 14.15*:

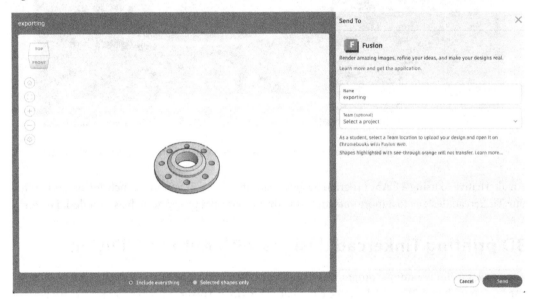

Figure 14.15: The options to send a model from Tinkercad to Fusion

In the **Send To** window shown in *Figure 14.15*, we have a few options to consider as we transfer our Tinkercad designs to Fusion. As with exporting a file, we can choose to send our entire design or only the parts that we have selected. We can also choose to rename the file, though this is optional.

If you are using the web-based version of Fusion, you will need to choose a **Team** to send the file to from within the web app. We can then press the **Send** button to send the file to either a project within the web app or the desktop version of Fusion if it is installed locally on your Windows or Mac device.

After pressing **Send**, we will be prompted to leave Tinkercad and switch to Fusion, as shown in *Figure 14.16*:

Figure 14.16: Prompt to switch between Tinkercad and Fusion

After pressing the **Open Autodesk Fusion 360** button that we can see in *Figure 14.16*, Fusion should automatically launch with our Tinkercad design loaded, as shown in *Figure 14.17*:

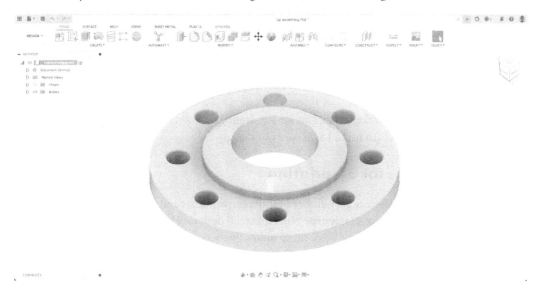

Figure 14.17: A Tinkercad design opened in Fusion

The window we first see, as shown in *Figure 14.17*, is Fusion's **DESIGN** space. Here, we could choose to use some of Fusions advanced modeling features to edit or manipulate our Tinkercad designs, as the design is completely editable, unlike when exporting an STL or OBJ file and bringing it into a traditional CAM program such as Cura. To switch between the different spaces and features in Fusion, we can press **DESIGN** to open the drop-down menu as shown in *Figure 14.18*:

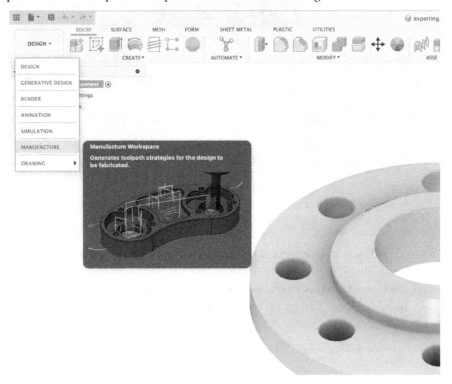

Figure 14.18: Navigating between windows in Fusion

If we just want to prepare our Tinkercad designs for 3D printing, we must switch to the **MANUFACTURE** window to access CAM tools to begin to prepare our designs.

Preparing designs for 3D printing

As shown in *Figure 14.18*, there are a number of different windows and features included in Fusion that could be used to enhance or manipulate our Tinkercad designs. However, the one we are looking for is the **MANUFACTURE** window. Once this window has been opened, we can prepare our designs for 3D printing through the following steps:

1. After opening the **MANUFACTURE** window, switch to the **ADDITIVE** tab and create a new setup by pressing the **SETUP** button, as shown in *Figure 14.19*:

3D printing Tinkercad designs with Autodesk Fusion   233

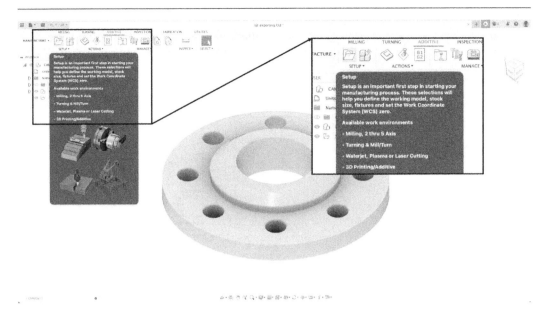

Figure 14.19: Pressing the SETUP button to start a new additive project in Fusion

2. In the **SETUP** window, we can press **Select…** to choose a machine. You can browse Fusion's library to see whether your 3D printer is included in Fusion's default profiles, or you can create your own custom printer. Fusion has an extensive collection of both extrusion-type and resin-type printers preloaded, as seen in *Figure 14.20*:

Figure 14.20: Browsing Fusion's machine library to choose a 3D printer

3. Once we've selected a 3D printer from the machine library, we can then press **Print Settings** to choose which material we want to use based on the printer we selected, as shown in *Figure 14.21*:

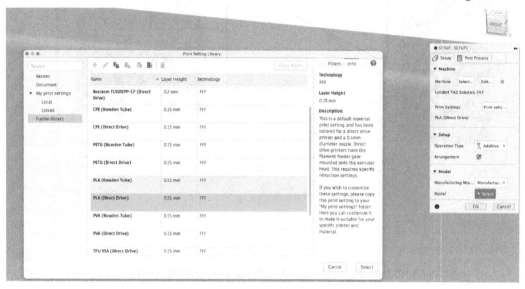

Figure 14.21: Browsing Fusion's print settings library

4. As we did previously in Cura, Fusion allows you to create and save your own custom print settings if you would like to adjust the resolution, infill, adhesion options, or settings for support material for your prints, as shown in *Figure 14.22*.

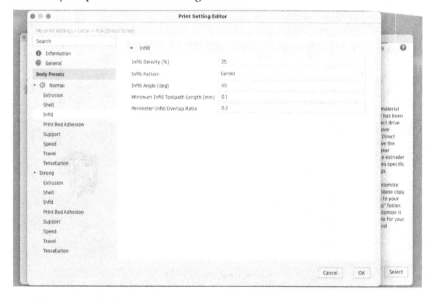

Figure 14.22: Adjusting and creating a custom print setting profile in Fusion

5. Once we've chosen our desired print settings, we need to ensure that the model(s) that we plan to print are selected in the **SETUP** window. To do this, click **Body** at the bottom of the **SETUP** window, then click on the models you want to include in the print so that they are highlighted, as shown in *Figure 14.23*. If you have parts in your design that you do not want to print, you can exclude them by not selecting them in this step.

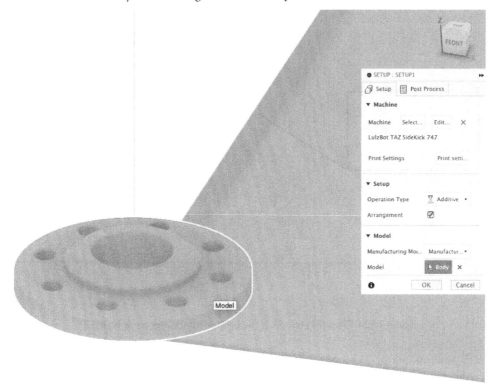

Figure 14.23: Selecting the parts in our design that are to be 3D printed

6. Once all parts have been selected, you can press **OK** to close the **SETUP** window. If you want to go back into the **SETUP** window and make adjustments, you can right-click on **Setup** and then select **Edit** in the **BROWSER** on the left side of the screen, as shown in *Figure 14.24*:

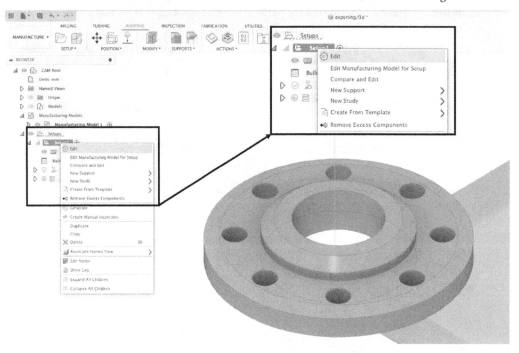

Figure 14.24: Choosing to adjust the print settings after exiting SETUP

7. If your model(s) are not located where you want them to be on the print bed, you can reposition them manually. You can also arrange multiple models to be spaced out across the print bed, though this is sometimes easier to do in Tinkercad, as discussed in *Chapter 11*. To do this in Fusion, press the move components button or *M* on your keyboard, then click and drag the arrows or type in dimensions to reposition a part as shown in *Figure 14.25*:

3D printing Tinkercad designs with Autodesk Fusion   237

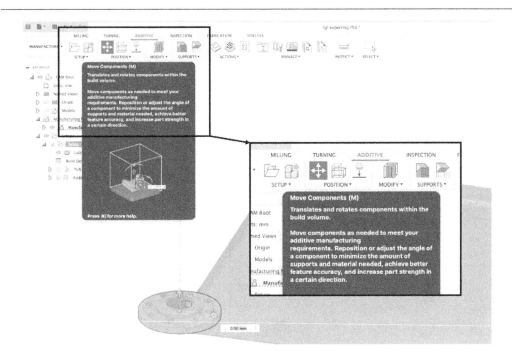

Figure 14.25: Adjusting the position of parts on the print bed

8. If a part is floating above the print bed, you can use the place parts on the platform tool to lower them so they are flat and touching the bed, as shown in *Figure 14.26*:

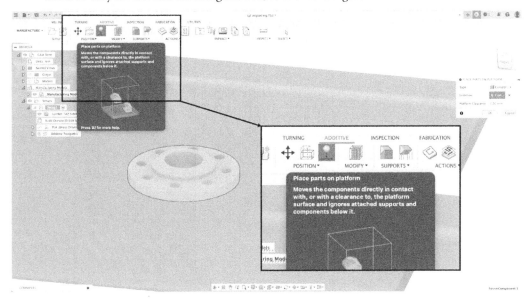

Figure 14.26: Adjusting models so they are touching the print bed

238 Exporting and Sharing Tinkercad Designs for Manufacturing

9. Once all settings have been chosen and models have been positioned, we need to slice our models as we did in Cura earlier on. In Fusion, this is called **Generate**, which can be done by pressing the **Generate** button on the toolbar, as shown in *Figure 14.27*:

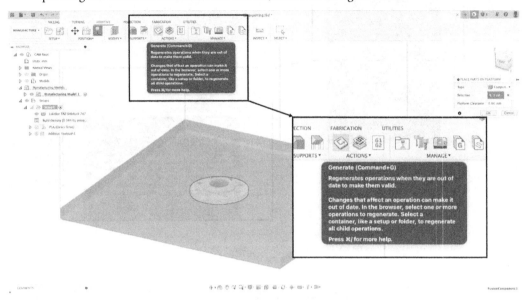

Figure 14.27: Generating the toolpaths for our prints

10. Once our settings have been generated, we can press the **Simulate Additive Toolpath** button on the top toolbar to open a preview, as shown in *Figure 14.28*:

3D printing Tinkercad designs with Autodesk Fusion   239

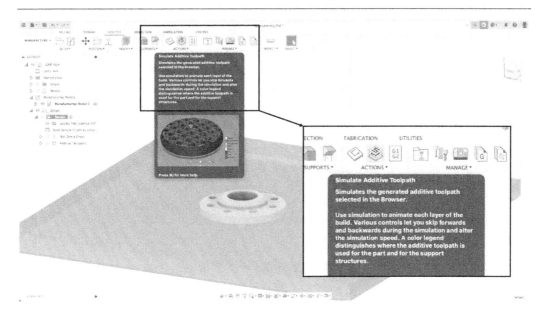

Figure 14.28: Simulating the 3D print in Fusion

Once in the preview simulation, we can scroll through the different layers, see how long this print will take, and check for any errors in the print, as shown in *Figure 14.29*:

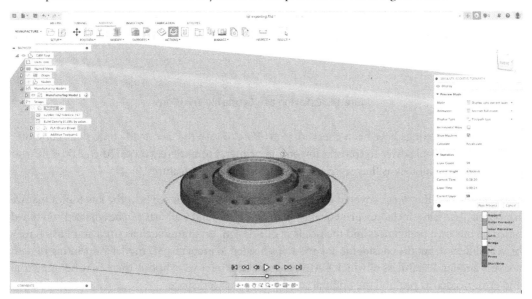

Figure 14.29: Previewing the 3D print in Fusion's toolpath simulation

11. Lastly, we can save our generated print settings as a Gcode file to start this 3D print. To do this, we need to press **Post Process** on the top toolbar to open the **NC Program** window. This is where you can change the name of the Gcode file, as well as check to make sure you are using the right settings for your printer, as shown in *Figure 14.30*:

Figure 14.30: Saving as a Gcode file in Fusion

12. Once all settings have been checked, press **Post** to save the Gcode file onto your computer or a removable drive to transfer to your 3d printer. From there, you can then 3D print your Tinkercad models as they were prepared in Fusion's CAM features!

Looking back on the topics covered thus far in this chapter, we can see how the key topics that we have been discussing, including print quality, material choice, and supports can be adjusted and tuned based on our own preferences and the needs of our design. Programs such as Cura and Fusion may be very different, but the considerations and steps needed to prepare a 3D model for printing in both are very similar. Regardless of which CAM program or printer you choose, there may be some trial and error needed as you find the optimal settings for your workspace.

However, what if you don't have access to a 3D printer, or if perhaps you need a different kind of printer for a project that you are working on? Fortunately, there are a number of services and steps you could take to have your models 3D printed and shipped right to your door!

Finding 3D printing services

A **3D printing service** is a company that has one or more print farms that you can use to print your own designs. There may be libraries or public maker spaces in your area which allow you to print your models using their resources, usually with some type of membership required. There are also 3D printing services available through the web which would manufacture your designs and send them right to your door.

To view these services in Tinkercad, press **Export** in the top-right corner of the screen to open Tinkercad's export window, as shown in *Figure 14.31*:

Figure 14.31: Looking at the 3D print options in Tinkercad

Once the export window has been opened, we can switch to the **3D Print** tab to see options for 3D printing our designs as shown in *Figure 14.31*. If you own a printer with cloud printing capabilities such as a Makerbot, you may be able to send your model directly to your printer's CAM software from this window. There are also apps, such as **Astroprint**, **3DprinterOS**, or **SimplyPrint** that allow you to make virtually any 3D printer cloud-print-enabled. It's important to note that there are other brand printers or cloud printing services available in addition to the ones listed here on Tinkercad, such as *Bambu Lab* or *Octoprint*, that would also be compatible with your Tinkercad designs. However, these built-in services add a bit of convenience by removing the need to export your model and upload it manually as you can just press one of these buttons to export and upload a design directly from Tinkercad.

However, there are also some services listed in this window, such as **Treatstock**, **i.materialise**, or **Print a Thing**, that will allow you to order your models to be printed. It's important to note that there are many other services available in addition to the ones listed here on Tinkercad. Some other popular services include *Shapeways*, *Protolabs*, and *Xometry*, and there are countless others as well.

As you are choosing a 3D printing service, it is important to do your research first. Read online reviews, check and compare pricing, ensure that they ship to your region, consider return or refund policies, and consider which service best suits your needs. For example, some of these services may only have FFF or SLA-type printers available, while other services may offer higher quality and industrial-type FDM or SLS printer services for you to use, though typically at a higher cost.

After choosing a service, you will be prompted to choose some default print settings, as we discussed earlier for CAM. This can include the type of printer, material or color, infill density, and print quality. All these settings will change the cost to produce your part, as shown in *Figure 14.32*:

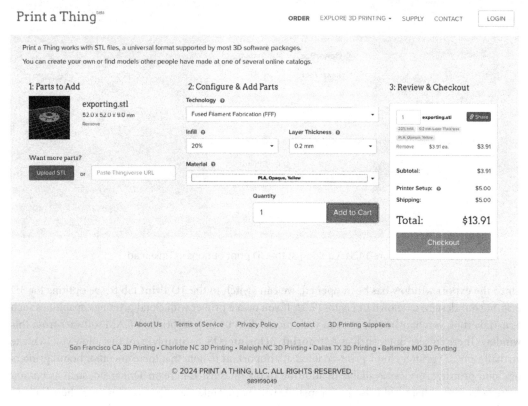

Figure 14.32: Choosing settings to order a 3D print (image credit: www.printathing.com)

As seen in *Figure 14.32*, a 3D printing service such as `printathing.com` can read your models from Tinkercad, and then let you choose your print settings to pay to manufacture your design. This is a great option for readers who do not personally own a 3D printer or have a local maker space available to them, or even readers who are looking to create something a bit more complex.

Even though I personally own a number of 3D printers in my maker space, I sometimes use a printing service when I need a part made using a technique or material that I don't have access to. Understanding the preparation steps needed, as well as how different print settings may affect the quality of your manufactured models, will allow you to make more informed decisions when choosing and using 3D printing services.

Summary

I have found that there is a common misconception that designing or choosing a model is the only step needed to be able to have a 3D-printed part. However, as discussed in this chapter, the design and modeling stages of the production process are really just the first steps.

Once we've created a model in Tinkercad, and created it using effective practices as discussed in earlier chapters, we can move to bring that model into a CAM program to prepare it for printing, as we learned in this chapter. We also learned that while there are many different CAM programs for 3D printing out there, the considerations and steps needed to prepare a model are often very similar.

Furthermore, we learned that we need to ensure that we choose a program that is compatible with our printer, as well as adjust the print settings to not only match what our printer can do in terms of material and quality but also support the needs of our model. If we are looking to make something quickly, we might choose a lower resolution than if we wanted to make a small and detailed part. We may also need to enable support material and adjust the infill density or pattern to ensure our designs are printed efficiently and effectively, as was mentioned in this chapter.

While CAM programs will offer default settings for us to use, we learned that it can sometimes take practice and experimentation to dial in custom settings that give us the most desired effect. We learned not to be discouraged when a print fails, and to always work to save your settings so you can reuse them for future success.

If you find yourself without a printer, or in need of a different type of printing technique, there are countless locations and services available that you can share and manufacture your models with too, as we learned in this chapter.

Now that we have learned how to use Tinkercad to design models for 3D printing effectively, as well as what 3D printers are and how they work, it's time to put these skills to the test through practical applications and challenges. Move on to the next part of this book when you're ready to get started!

Part 4: Practical Applications, Start to Finish Designs to Test our Skills

Unlike the first three parts of this book, the fourth and final part focuses more on specific projects rather than building general skills and understanding. Each chapter in this part allows us to apply the skills and concepts learned in previous chapters through practical applications for designing models in Tinkercad, and then manufacturing those models successfully with 3D printing. We will not only continue to build proficiency in using Tinkercad through advanced techniques, but also gain a greater understanding of the different considerations and resources you might need when striving to manufacture your projects most effectively. We will also emphasize the design process as we reflect on what we make, as well as how to improve as mistakes are made while we continue to learn and grow in our abilities.

This part includes the following chapters:

- *Chapter 15, Designing and Printing a Trophy*
- *Chapter 16, Fabricating a Multi-Part Storage Box with a Sliding Lid*
- *Chapter 17, Modeling an Ergonomic Threaded Jar*
- *Chapter 18, Building and Playing a 3D Puzzle*
- *Chapter 19, Designing and Assembling a Catapult*
- *Chapter 20, Prototyping a 3D-Printed Phone Case*

15
Designing and Printing a Trophy

As we've entered our fourth and final part of this book, the following chapters will be a test of our skills through practical applications for designing and manufacturing our Tinkercad models with 3D printers! Each practical application will allow us to apply the previously learned topics and skills covered throughout the book, as well as explore real-world challenges and activities that put these skills to the test.

In this chapter, we will be celebrating our newly acquired skills and abilities by designing and printing a trophy! The steps to achieve this will be covered in the following topics:

- Designing the top part
- Designing the base part
- Reorienting our parts
- Exporting and preparing for production
- Manufacturing the models

As we progress through each of these topics, we will strive to design an aesthetically pleasing trophy by combining different shapes through constructive solid geometry. Later, we will look at how our models can be exported, prepared, and manufactured through 3D printing production techniques specific to the needs of this project as we strive for success!

Technical requirements

To successfully design and manufacture a 3D-printed trophy, you will need access to the Tinkercad design program to be able to design your models using a web-enabled device. As discussed earlier in *Chapter 2*, I also recommend that you utilize a mouse to aid in designing complex models more effectively.

To then manufacture your design, you will need access to a 3D printer, whether it be your own or whether you access it through a 3D printing production service as discussed in *Chapter 14*. We will be looking at settings to print our trophy effectively in **CAM**, which will be needed to accompany your 3D printer as we also discussed in the previous chapter.

You will need **3D printing material** to print your trophy with, whether it be filament for an extrusion-type printer or resin for a vat photopolymerization-type printer. For novelty projects such as this trophy, I like to use specialty filaments that add to the aesthetics, such as metallic silk PLA.

Lastly, the example trophy model created throughout this chapter can be found at this link: `https://www.tinkercad.com/things/gtpWFXcatTz-trophy-model-from-chapter-15`

Designing the top part

The first part of our trophy is going to be the top half, which will stand prominently atop a base that we will be designing later in the *Designing the base part* section. We will break this first part down into a few steps, starting with the cup itself.

Modeling the cup

To model the cup, we will follow these steps:

1. After creating a new Tinkercad 3D design, grab a **Paraboloid** shape and bring it onto your workplane. We can then dimension this shape to be however large we want our trophy to be. As an example, I have modeled mine to be 85 x 85 x 65 mm in L x W x H as shown in *Figure 15.1*:

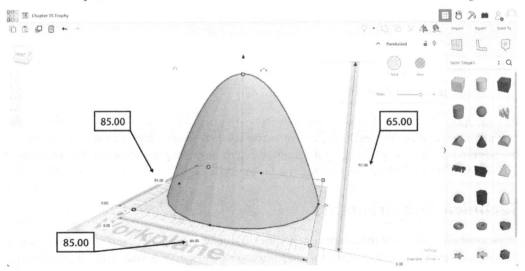

Figure 15.1: Dimensioning a Paraboloid for the cup

2. Next, we can rotate the **Paraboloid** by 180 degrees so the flat surface is facing up, as shown in *Figure 15.2*:

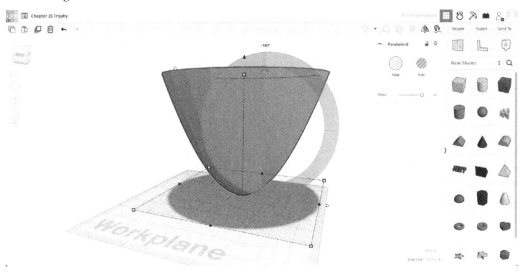

Figure 15.2: Rotating the paraboloid

3. Now we are going to make a copy of the **Paraboloid**, and then scale this copy down so it fits within the original one. We can then raise the new **Paraboloid** up off the workplane by 5 mm, and set it to be a **Hole**, as shown in *Figure 15.3*:

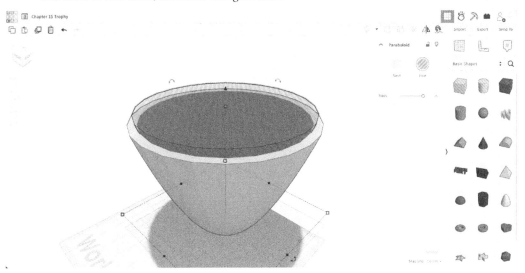

Figure 15.3: Creating an inner-hole paraboloid

4. Next, select both paraboloids and raise them off the workplane by 55 mm, as shown in *Figure 15.4*.

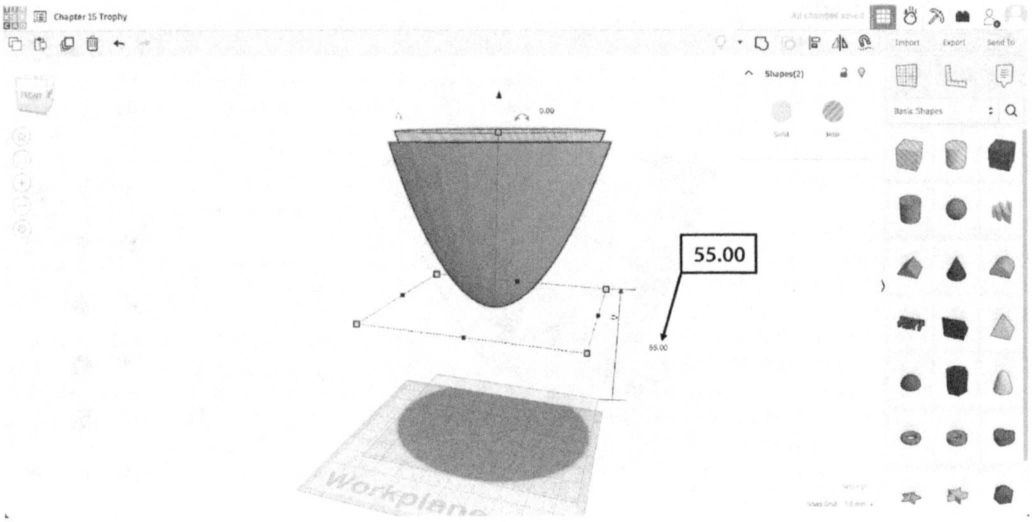

Figure 15.4: Raising the Paraboloid shapes

Raising the paraboloids like this will leave room below them to create the post that the cup shape rests on, which will be created in the next section.

Modeling the post

To model the post, we will follow these steps:

1. We are now going to bring a **Cylinder** shape into our design and place it below the raised paraboloids. It doesn't need to be perfectly centered just yet, but we do want to dimension it to be 15 x 15 x 65 mm and with 64 **Sides** from the shape's parameter window, as shown in *Figure 15.5*:

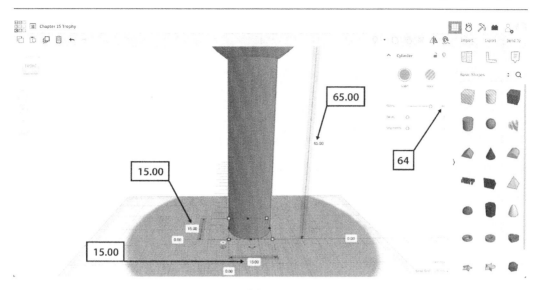

Figure 15.5: Modeling a cylinder as a post

2. Next, we are going to bring a **Tube** shape into our model as this shape allows for more adjustments in the shape parameters than a cylinder. We can dimension the **Tube** as 50 x 50 x 10 mm and give it a wall thickness of 10, 64 **Sides**, and a larger **Bevel** as shown in *Figure 15.6*. The **Bevel** and overall size of this **Tube** are not critical, so choose dimensions that look aesthetically pleasing to you.

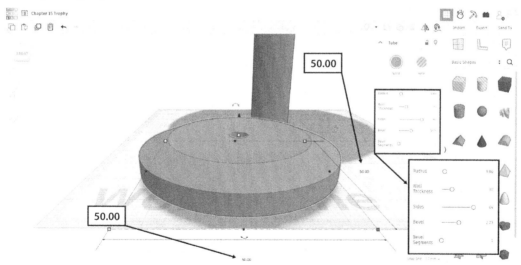

Figure 15.6: Creating a tube shape

3. We can then move the **Tube** so it is below the paraboloids and cylinder as shown in *Figure 15.7*, though we don't need to align these shapes perfectly just yet.

Figure 15.7: Repositioning the tube

4. Now, to add to the aesthetics a bit, let's bring in a **Torus**, then scale and place it over the base of the post as shown in *Figure 15.8*. Again, don't worry about aligning everything just yet.

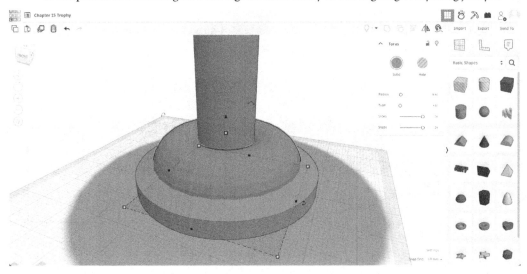

Figure 15.8: Adding a torus shape

5. Next, we can either make a copy of the **Torus** or bring in a second one. After that, let's scale the new **Torus** down and raise it higher off the workplane, as shown in *Figure 15.9*. The specific dimensions of these **Torus** shapes are not critical. Arrange them so that you like the appearance.

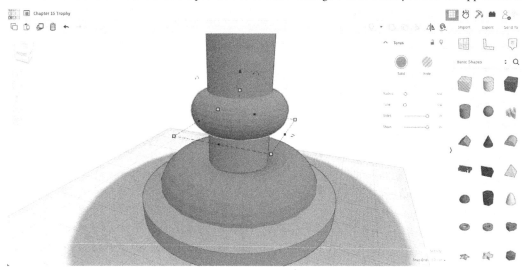

Figure 15.9: Arranging a second torus

6. In *Chapter 7*, we learned how to use the **Duplicate and repeat** tool to copy and move shapes to make patterns in our designs. Let's use this tool now to copy the torus two more times and create a pattern up the cylinder post as shown in *Figure 15.10*.

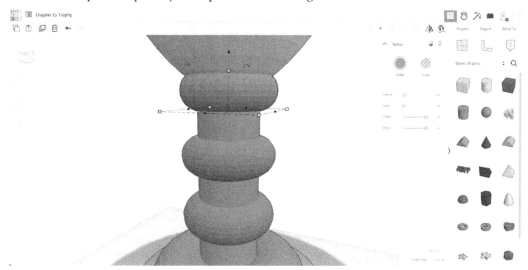

Figure 15.10: Creating a patterned post with torus shapes

Now that we've created a post below the cup part, we can move into finishing that part of the trophy and getting all of the shapes aligned and grouped perfectly in the next section.

Finishing the cup part

To finish the cup part of the trophy, we are going to follow these steps:

1. Bring in another **Torus** shape and rotate it by 90 degrees so it is vertical. We can also dimension this torus so that it is approximately 35 mm in diameter, though the dimensions for this are not critical as this too is an aesthetic part, as shown in *Figure 15.11*:

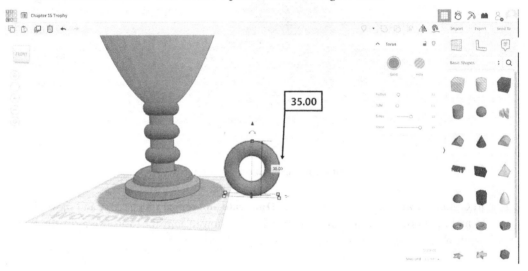

Figure 15.11: Creating a vertical torus shape

2. We can then raise the **Torus** up so it is approximately 80 mm above the workplane, then move the shape into the cup we made earlier to start to create a handle, as shown in *Figure 15.12*:

Designing the top part

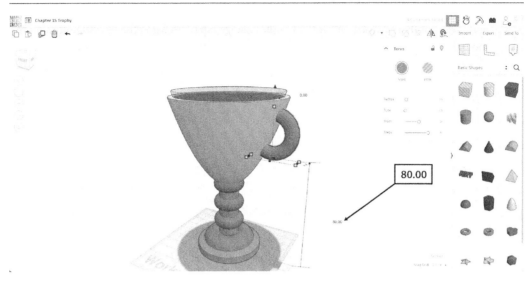

Figure 15.12: Starting the handle design

3. Next, make a copy of this raised **Torus** and bring it to the opposite side of the paraboloids. We don't need to be concerned with exact positions just yet. Instead, work to bring the shape into an approximate location as shown in *Figure 15.13*:

Figure 15.13: Adding a second torus for the handles

4. Now select both **Torus** shapes for the handles and group them using the **Group** button on the toolbar, or by pressing *Ctrl + G* as shown in *Figure 15.14*. This will allow us to easily align our shapes in the next step.

Figure 15.14: Grouping the handle shapes

5. Now that our shapes are partially arranged and grouped, we can use the **Align** tool to perfectly center the paraboloid, cylinder, tube, and torus shapes as was initially introduced in *Chapter 7*. First, select all shapes, then press the **Align** button on the toolbar, or press *L* on your keyboard. We can then use the dots shown in *Figure 15.15* to align our shapes across the *x* and *y* axes on the workplane:

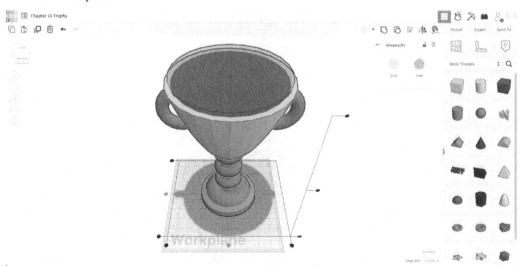

Figure 15.15: Aligning all of the shapes for the cup part

6. Lastly, we can **Group** all shapes together while they are still selected from the previous step. Once grouped, we can adjust the color as desired and review our model for the top part of the trophy project as shown in *Figure 15.16*:

Figure 15.16: Grouping shapes to make the trophy cup

Now that we've completed the top part of our trophy, we can move this model off to the side and begin to work on the next part: the base.

Designing the base part

With the top part of our trophy model completed, we are going to move into designing a base for the cup and post to rest on. This could be designed to be a single part, with the top part we designed previously connected permanently to the base, or the base could be a separate part that is assembled later. By modeling these parts separately, you can easily export and print them in different materials as demonstrated later in the *Exporting and preparing for production* section, though this is not a requirement for success. Regardless of which direction you choose to go in, we will start in a similar way as described through the following steps.

Creating a base platform

To create a base platform, follow these steps:

1. After moving the cup part off to the side, we can bring a new **Cylinder** shape into our design. This cylinder can be dimensioned to be 85 x 75 x 10 mm so that it is an oval, and with 64 **Sides** so that the edges are smooth as shown in *Figure 15.17*:

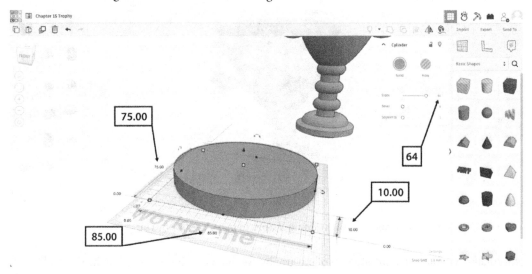

Figure 15.17: Creating a cylinder for the base

2. Instead of using the bevel parameter as we did in the *Modeling the post* section earlier, we are going to add a **Cone** shape on top of the cylinder that we created in *step 1* to add a chamfered edge to the top of our base. The **Cone** should be dimensioned to be the same in length and width as the cylinder, and the height of the **Cone** is not critical. We can adjust the height of the **Cone** to manipulate the size of the chamfer angle that we will be creating later in this section, as shown in *Figure 15.18*:

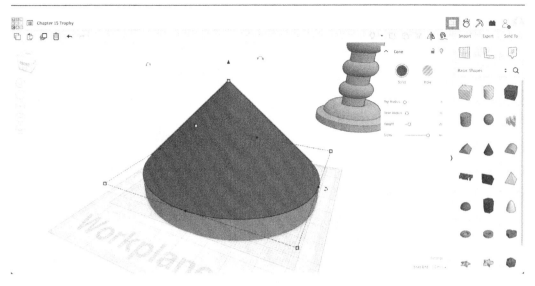

Figure 15.18: Adding a cone to the base

3. We can then add a **Box** shape set as a **Hole** and scale it so it is larger than our base. This **Box** can then be raised off the workplane by about 12 mm so that it covers most of the cone that we created previously as shown in *Figure 15.19*:

Figure 15.19: Adding a box hole shape to cut the base

4. Now, select the cylinder, cone, and box shapes at the same time and **Group** these shapes together to create a solid-based shape with a chamfered edge as shown in *Figure 15.20*. If you want to make the chamfer larger or smaller, you can ungroup the base, then raise or lower the box shape before regrouping.

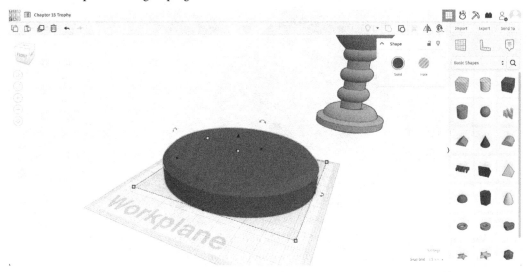

Figure 15.20: Creating a chamfered base

With the initial part of the base created, we can move into designing a spot for the cup part and base part to connect.

Creating a connection point

We are now going to create a point where the cup part that we made previously will connect to the base part we've started working on. You could choose to make this a solid connection point, or one that is separate so that the parts can be printed separately as discussed earlier on. If you want two separate parts, here's how you would do that:

1. Make a copy of the cup part and bring it onto the base so it is in position. I have chosen to center the trophy on the *x* axis, but pushed toward the back of the base, as shown in *Figure 15.21*:

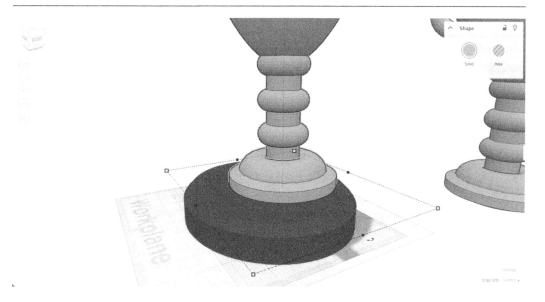

Figure 15.21: Placing a copy of the cup part on the base

2. We can then turn the copy of the cup part into a **Hole** shape, and then use it to create a cut out of the base by grouping the hole shape to the base shape, as shown in *Figure 15.22*. This cutout will match the bottom of our cup part perfectly, making assembly seamless later if we choose to design these as separate parts.

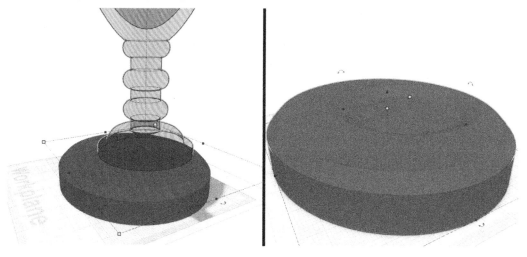

Figure 15.22: Creating a cutout in the base for the trophy parts

This has created a nice place to attach and glue the two parts together during the post-production stages, as we will look at in the *Manufacturing our models* section later on. However, you could also simply group the cup to the base as a solid instead of a hole to make this a single part if that's what you prefer.

Adding the recipient

The last step in completing the base part is to add the recipient who has earned this trophy! We can do this in many ways, and I like to utilize the different text generators that we looked at earlier in *Chapter 5*, as shown in the following steps:

1. Search for and find the **curved words** shape from the **Community** shapes library and drag this shape onto the **Workplane** as shown in *Figure 15.23*:

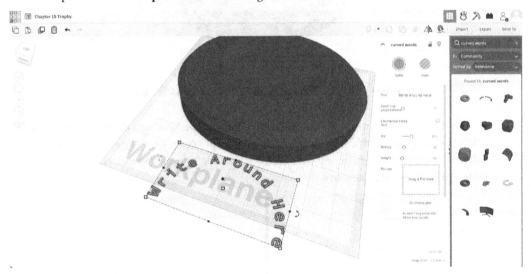

Figure 15.23: Adding the curved words shape to our design

2. Next, lift the words so they are on top of the base, then adjust the shape parameters to include the recipient for this trophy and adjust the parameters so that the words curve nicely around the front of the base. These measurements are not critical, but they are aesthetic. We are adjusting the text according to our preferences as shown in *Figure 15.24*:

Figure 15.24: Adjusting the curved words shape's parameters

3. Lastly, we need to group the words with our base to make a solid part. There are two different ways we could do this depending on our preferences, as shown here:

   A. We could group the words with the base as solid shapes so that they extrude from the surface of the base, as shown in *Figure 15.25*. While this offers a prominent raised text feature, you may find that this is a more difficult 3D print to achieve successfully when compared to the alternative shown in *option B*. This is because raised text such as this needs to be filled in, and when the corners of the text are quite small, the nozzles of extrusion-type 3D printers may struggle to fill in the shape accurately.

Figure 15.25: Grouping the text as a solid shape

B. As an alternative, we could set the text to be a hole shape, then group this shape so that it creates a pocketed text feature in the base, as shown in *Figure 15.26*. This offers a different aesthetic to the one shown in *option A*, and one that might be easier to manufacture on a lower resolution 3D printer, as the nozzle can now extrude a larger amount of material around the text rather than needing to be able to extrude the detail of each individual letter. However, choosing between these options really comes down to personal preference when designing a novelty item such as this.

Figure 15.26: Grouping the text as a hole shape

Now that we've completed the base, we can move into reorienting the parts of our design as we make some final adjustments.

Reorienting our parts

Since we created the cup and base parts separately, we haven't had a chance to review our trophy design as a completed model. Let's adjust the color of our parts and bring them together as shown in *Figure 15.27*:

Reorienting our parts 265

Figure 15.27: Assembling our parts

Bringing parts together in Tinkercad allows us to prototype our designs and inspect them for any possible problems that may arise later. We can also make copies of our designs and experiment with different iterations easily, as discussed earlier in *Chapter 12*.

With aesthetic and novelty projects such as this trophy, the specific dimensions are often less important than the scale and overall proportions. This allows us to scale our designs to make different versions, such as a mini trophy, as shown in *Figure 15.28*:

Figure 15.28: Making a smaller-scale trophy model

However, before we move into exporting our designs for manufacturing, it's important that we reorient the parts to prepare them for manufacturing. This includes separating them if we plan to print the parts using different materials, as well as ensuring that all models are touching the **Workplane** and arranged for 3D printing production as discussed in *Chapter 11*. To do this, let's remove the cup part from the base and lower it so it is also touching the **Workplane** as shown in *Figure 15.29*:

Figure 15.29: Separating parts and preparing them to be exported

With our parts separated and touching the **Workplane**, we are ready to move into exporting the models and preparing them for production in our CAM software.

Exporting and preparing for production

At this stage, it is time to take our models out of Tinkercad to prepare them for production with 3D printing techniques. To do this, we must first select a part of our design, then press the **Export** button in the top-right corner of our design window. This will open the **Download** window, as seen in *Figure 15.30*:

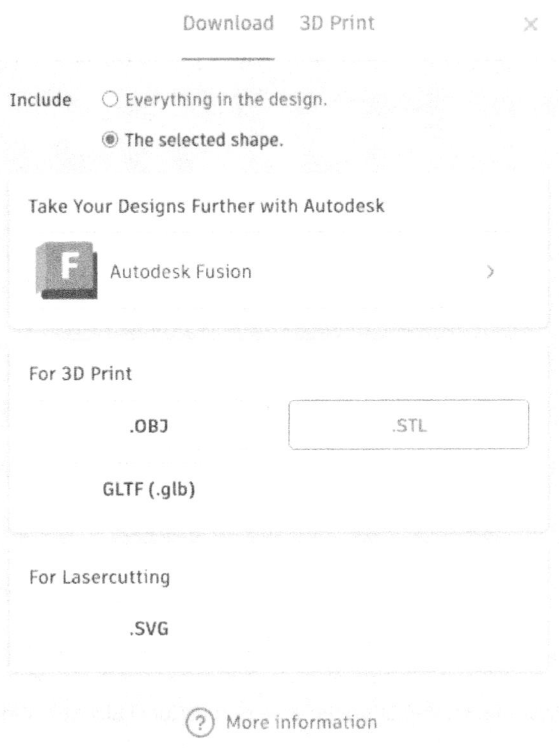

Figure 15.30: Options for downloading our models

If you are choosing to 3D print this project yourself, you will want to download it using one of the **For 3D Print** options, as discussed in *Chapter 14*. If you are printing the base and cup part separately using two different materials, then you would repeat this process for each part. Alternatively, you could switch to the **3D Print** options rather than the **Download** options so that you could order these parts to be manufactured via a 3D printing service that we also looked at in the previous chapter.

If you are producing these models yourself, then you need to bring the exported files into your CAM program for your 3D printer. As discussed in *Chapter 14*, there are a lot of different slicer programs out there. You should either choose the one that is recommended by your 3D printer's manufacturer, or a generic one that suits your own needs. In the following steps, I will be demonstrating how to prepare these trophy models using *Cura LulzBot Edition*, though similar steps can be taken with whichever CAM program you are using.

As I will be printing the cup part and base part separately with two different materials, I will bring each exported file into my slicer program one at a time, starting with the base, as shown in *Figure 15.31*:

Figure 15.31: Preparing the base part in Cura LulzBot Edition

As learned in *Chapter 14*, there are a lot of settings we could consider, tweak, and change as we prepare our 3D models for production CAM software. However, printing these trophy parts should be straightforward with just a few important considerations:

- **Material**: We need to choose the correct material settings, which will adjust our print temperatures and speeds based on the selected material. I will be using a Metallic Silk PLA for both parts, which can be printed using the generic PLA profiles.

- **Quality**: This will determine our resolution or layer height, as well as the amount of material used and the overall printing time. I am going to print these parts at a higher quality, as I want a nice finish across my metallic trophy parts.

- **Infill Density**: This will determine how hollow the parts are, and for the base, we can reduce this to as little as 10% to save print time and cut down on material without having to worry about any structural issues.

- **Supports**: As the base is a flat part that is touching the build platform with no overhangs, no supports will be needed for this part, though I will leave a **Skirt** enabled to print around the perimeter of my model.

After adjusting the settings based on your printer, material, and desired finish, you can slice the model and preview it for inspection. Always remember to preview the first layer to ensure that the model is touching the build platform as intended. You can see the first layer of the base part in the preview in *Figure 15.32*:

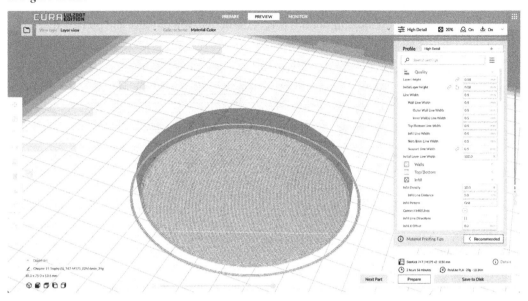

Figure 15.32: Previewing the first layer of the base part in Cura LulzBot Edition

Once you've inspected this part for any possible issues, you can save the **Gcode** file before moving on to the next part of the design.

Once you have brought the model for the cup part into your CAM program, you can adjust the print settings to suit the needs of this model. Like the base part, there are a few considerations to help achieve greater success:

- **Material**: I will be selecting PLA as my printing material again, though you might utilize two completely different materials as you've exported and prepared your models separately.
- **Quality**: I will be choosing a higher detail for the cup part again as I want to capture all the details of the design that we created earlier.
- **Infill Density**: Unlike the base part, the cup part has more overhangs and thinner wall areas to work around. As a result, I would keep the infill density at a minimum of 20% to avoid any possible failures.

- **Supports**: With the cup part, there are several overhangs to work around. We've designed these parts using rounded edges and shallow corners, which should make printing the overhangs more efficient. However, I am still enabling supports, specifically **tree supports**, so that the bigger overhangs such as the handles can be fully supported when they are printed.

Once you've adjusted your settings to suit this part of the design, you can slice and preview the model as shown in *Figure 15.33*:

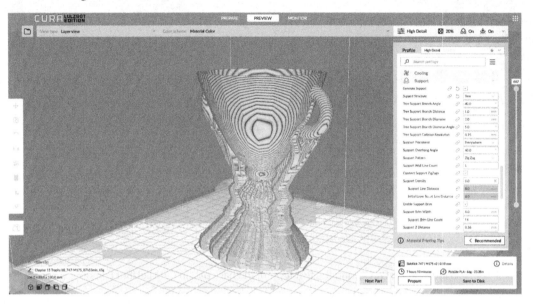

Figure 15.33: Previewing the cup part in Cura LulzBot Edition

After inspecting the cup part for any possible issues as well as looking at the first layer to ensure that our model is fully on the build platform, you can save this part as a *Gcode* file for production. From there, all that's left is to transfer these files to your 3D printer and move into the manufacturing stages of the project!

Manufacturing the models

Now that we've created our designs in Tinkercad and prepared them with the specific settings for our printer in CAM, we can move into manufacturing our models using our 3D printers! Setting up the material and starting the prints will vary a bit from printer to printer, but then the next step is to print your designs. *Figure 15.34* shows the printing of the base part using a FFF type 3D printer and Black Metallic Silk PLA:

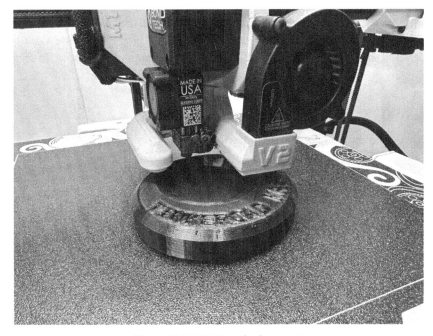

Figure 15.34: Printing the base part

Figure 15.35 shows the printing of the cup part using a FFF type 3D printer and Gold Metallic Silk PLA:

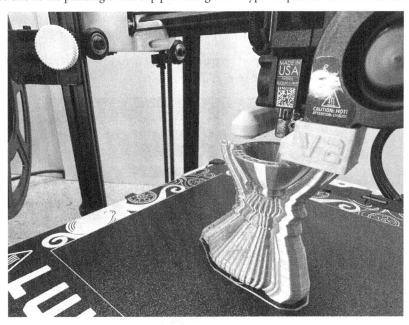

Figure 15.35: Printing the cup part

Once the prints finish (or arrive if you've utilized a printing service), we can move into post-production and assembly. Any supports will need to be carefully removed first. I like to use a combination of pointed pliers and small wire cutters to pull and trim away any supports, as shown in *Figure 15.36*:

Figure 15.36: Removing tree supports from the cup part

It's important to do this carefully as smaller parts in your design may also break away by mistake. You might find that some light sanding is also needed to get a perfectly smooth finish, though this may also scratch and scuff the surface when using a specialty material such as the Metallic Silk PLA I chose for this project. As discussed in *Chapter 11*, avoiding supports when possible is always best when looking for the cleanest finish, and choosing the right type of supports for the needs of your project is key.

As I have printed the cup part and base parts separately, I also need to assemble these parts to create the final project using glue, as shown in *Figure 15.37*:

Figure 15.37: Applying glue to assemble the 3D-printed parts

Choosing the right glue or adhesive is an important step in finding success, and one that will change based on the materials that you are working with. I find that generic superglue works well for most applications, such as gluing the parts of our trophy together. There is also specialty epoxy for working with different types of materials, such as PLA, as well as silicon adhesives for bonding flexible filament parts together. A good strategy is to do your research and see what the manufacturer of your print material recommends and to do some sample test pieces before diving into a big project.

In the next few chapters, we are going to look at making multi-part models that assemble without the need for adhesives as there are a lot of different ways to create a successful multi-part project! However, once you've gotten your parts together, you should be rewarded with a stunning trophy such as the one shown in *Figure 15.38*:

Figure 15.38: The assembled trophy project

Well, done! You have not only designed and created a real-world project using Tinkercad and 3D printing technology but also successfully put some complex skills and strategies into practice. As you continue to progress through the remaining chapters of this book, we will continue to increase the complexity of our practical applications as we review and apply the various skills and topics that we've been covering.

Summary

Looking back at the challenges and activities covered throughout this chapter, there are a few key things that we should recap before moving on to the next one.

First, we learned that it is important to consider all aspects of the production process throughout the design phases, and not only toward the end when we are getting ready to manufacture our models. As you design a model, even a novelty item such as a trophy, consider the size of your printer, the resolution and quality it can achieve, and the materials and resources you have available. There is never one way to do any project; choose constraints and specifications that cater to your individualized needs.

As we design models such as the trophy created throughout this chapter, consider that some dimensions may not be critical, as was the case in this chapter. Instead, have fun with Tinkercad's design features to make aesthetic features that look and perform well without being too concerned about minor details. We can't always get away with this, as sometimes the level of precision and tolerance needed to find success is higher, but in this chapter, we could.

The CAD and design stages are just the first step toward finding success. The greatest model in the world cannot be made successfully if we don't also choose the right resources and get our print settings correct to match, as we learned in this chapter. Remember to test and prototype your designs so you can find the optimal settings as you continue to engage with these concepts.

When you're ready to look at another multi-part project, and one that doesn't need glue but instead tighter tolerances and greater precision, turn to the next chapter to dive into our next practical application!

16
Fabricating a Multi-Part Storage Box with a Sliding Lid

In this chapter, we will continue to apply our skills and abilities for designing models in Tinkercad that are manufacturable with 3D printing technology through another practical application! For this project, we will be creating a multi-part storage box with a sliding lid through the following topics:

- Starting with the box
- Cutting the grooves
- Making the lid
- Adding artwork
- Exporting and manufacturing our models

As with the last chapter, this practical application will allow us to recap and review key concepts covered earlier in this book as we design a prototype under the constraints of a real-world challenge. This project will have a greater emphasis on dimensions and measurements as initially introduced in *Chapter 2*, as well as tolerances as discussed in *Chapter 12*.

By combining these concepts successfully with good practices and strategies for 3D modeling and 3D printing, we will be able to manufacture a project that is not only aesthetically pleasing but useful too!

Technical requirements

In addition to the software that is needed to design and prepare your models for production, you will also need access to a 3D printer to manufacture your projects. To manufacture the box that we are making in this chapter, a wide range of rigid materials could be used for 3D printing, such as PLA, PETG, or ABS, as discussed in *Chapter 10*.

You may also want to incorporate graphics or images into your design as we will be discussing later in this chapter. As initially introduced in *Chapter 8*, you can use programs such as *Google Drawings* or *Vectr* to make your own images that can be imported, or you can browse image repositories such as *SVG Repo* to find ones to use in your design.

Lastly, the example box model created throughout this chapter can be found at this link: `https://www.tinkercad.com/things/djf0u9yccot-box-model-from-chapter-16`

Starting with the box

The first part of this project is to design the outermost part of the storage box. To do this successfully, we must first decide what our storage box will hold. For example, are you planning to hold pencils or tools? Perhaps makeup or jewelry? Or maybe use this box to hold bolts and fasteners for your workshop? This last option is what I have chosen to do.

After you've decided what you want your box to store, take some rough measurements to use in your design and jot them down on a Post-it or in a simple sketch to plan your project. We first looked at how to take measurements and plan projects effectively in *Chapter 2*, and this is a key step to finding greater success. A pencil case wouldn't perform very well if a pencil couldn't fit in it, after all! It's also important that you keep your production resources in mind, too. For example, a 3D printer with a 150 mm print bed wouldn't be able to make a box that is 200 mm in length.

Once you've decided how big your box needs to be, we can begin modeling the first part of this project through the following steps:

1. After creating a new Tinkercad 3D design, drag and drop the **Box** shape onto the workplane, as shown in *Figure 16.1*:

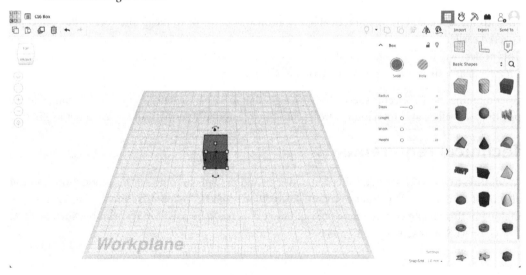

Figure 16.1: Starting with a Box shape

2. We can then set the dimensions of this shape to be big enough to store what we've chosen. We want to add a bit of length and width to the measurements we took earlier to account for some wiggle room, as well as to account for the wall thickness. For example, I would like a storage space of 140 mm x 40 mm x 25 mm in L x W x H, so I have dimensioned my box to be 10 mm bigger all around, as shown in *Figure 16.2*:

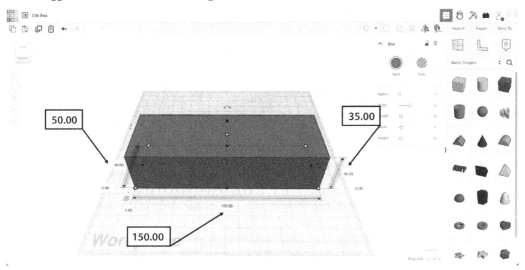

Figure 16.2: Dimensioning the Box shape

3. Next, we can make a copy of the **Box** shape and dimension the copy to be our inner pocket where our items will be stored. The copy should be set to be a **Hole** shape rather than **Solid**, and it should be smaller than the first box we created. The difference in size will become the thickness of the walls for our box. Thicker walls would produce a stronger box, but also use more material and take longer to 3D print. I've made the copy 5 mm smaller in length and width, which will give me 2.5-mm-thick walls. I've also raised the copy 2.5 mm above the workplane to create a 2.5-mm-thick floor, as shown in *Figure 16.3*:

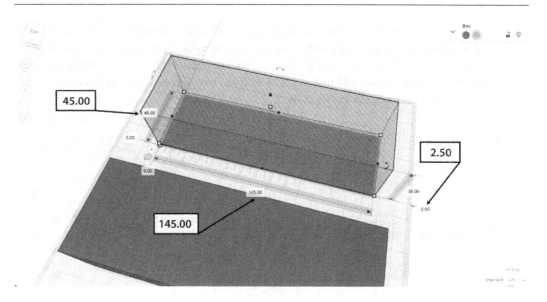

Figure 16.3: Creating and dimensioning the storage cavity for the box

4. Next, we can move the **Hole Box** shape into the center of the **Solid Box** shape by using the **Align** tool. To use this tool, select both shapes and press the **Align** button on the toolbar, or *L* on your keyboard. Then, press the *center dots* to align in the *X* and *Y* axes, but do not align the *Z* axis so that the hole shape cuts through the top of the box and leaves a floor below, as shown in *Figure 16.4*:

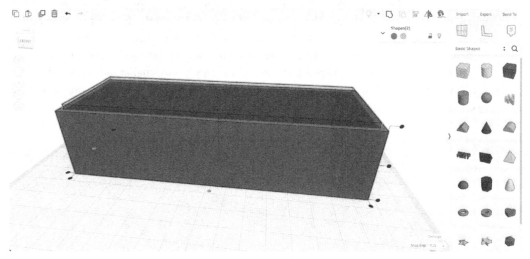

Figure 16.4: Aligning the two Box shapes

5. Lastly, we can finish this part of the project by grouping the two shapes together to create the outer box. With both shapes selected, press the **Group** button on the toolbar, or *Ctrl + G* on your keyboard, to cut the hole out of the solid, as shown in *Figure 16.5*:

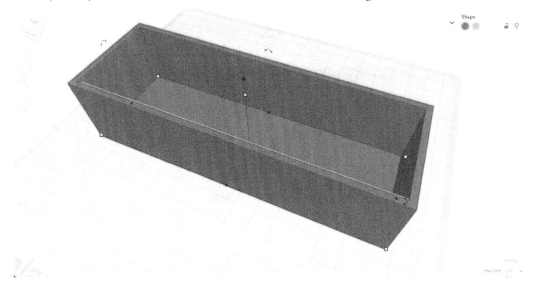

Figure 16.5: Grouping the two Box shapes

With this part of the project completed, we can move on to the next part, which are the grooves that allow for the lid to slide open and closed.

Cutting the grooves

The grooves are going to serve as tracks that allow for the lid to slide for the box to open and close securely. The structure of these grooves will be an *overhang*, and as discussed in *Chapter 11*, creating overhangs often requires the need for *support material*. However, placing support material in these grooves may make a rougher surface which would reduce the functionality of the box. Fortunately, the grooves will be very small with a minimum overhang that should be manufacturable without needing any support.

Before we model the grooves, let's change our viewing angle so that we are looking at the right side of the box from a 2D *orthographic* view, as shown in *Figure 16.6*:

Figure 16.6: Looking at the side of the box from an orthographic perspective view

This view will allow us to model the grooves more effectively as it focuses our perspective on one side of the box. As discussed in *Chapter 3*, changing our views is an important step in mastering Tinkercad's 3D design space.

Once we've repositioned our screen, we can begin to model the grooves through these steps:

1. First, let's drag and drop a new **Box** shape in our design and set it to be a **Hole** shape. We then want to position this hole shape so that it is flush with the top of our solid box and the same width as the opening, which in my case is 45 mm. We also can set the height of this box to be how thick we want the lid to be, which I have set to be 5 mm. Your measurements might vary based on the size you've chosen for your box, but the positioning should look as mine does in *Figure 16.7*:

Figure 16.7: Creating the first part of the grooves

2. We are now going to create a second box that will become the grooves, or tracks, for the lid to slide in. To do this, use the **Duplicate and repeat** tool to make a copy of the **Hole Box** shape we made in the last step. Then, dimension this copy so that it is 2.5 mm wider, only 2 mm tall, and flush with the bottom of the first box, as shown in *Figure 16.8*:

Figure 16.8: Modeling the grooves

By extending the width of this box by just 2.5 mm, we will be able to make small grooves for the lid to slide in while also keeping the overhangs to a minimum, which should allow us to avoid needing support material during the manufacturing stages.

3. We can now change our view back to being a 3D perspective view so that we can look at the entire box. We can then drag the hole shapes made for the grooves so that they extend from the right side of the box to the inner back wall to expand the tracks, as shown in *Figure 16.9*:

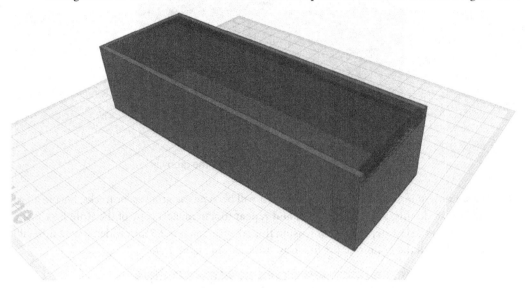

Figure 16.9: Expanding the groove tracks along the opening of the box

4. Lastly, we are going to group the two **Hole Box** shapes with the **Solid Box** shape to cut the grooves. But before we do, let's make a copy of the **Hole Box** shapes and move them off to the side to use in a later step, as shown in *image A* of *Figure 16.10*, then group the solid box with the original hole boxes, as shown in *image B* of *Figure 16.10*:

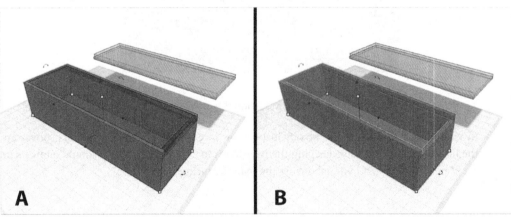

Figure 16.10: Copying the hole shapes and grouping the shapes to make grooves

With the grooves cut, we can move on to modeling the last part of the box, the lid.

Making the lid

We are now going to model the lid for our box, which will be able to slide in the grooves that we created in the previous section. To start, move the copy of the groove boxes back into the main box part, then set the copy of the groove boxes to be **Solid** shapes for the lid. I also chose to make the lid boxes a different color than the other box, so that they are easier to see, as shown in *Figure 16.11*:

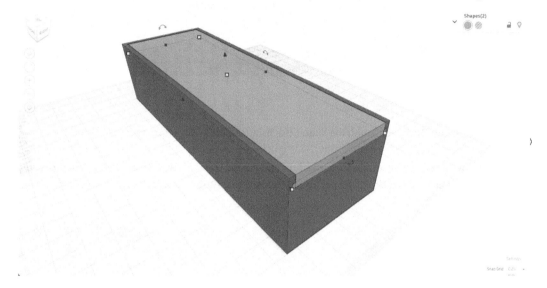

Figure 16.11: Starting the lid using existing shapes

We will use these boxes as a starting point to model the lid, which will need to be smaller to fit and slide in the grooves. As mentioned in *Chapter 12*, creating *wiggle room* between parts that need to fit together are called **tolerances**. To do this step successfully, you will need to identify the tolerances for your 3D printer and also for the material that you are using, also as discussed in *Chapter 12*.

We need the lid to slide freely, so we are going to create a **clearance fit**. For my resources, I have identified that a tolerance of 1 mm works well to create this type of fit, which I will be using in the following steps to model the lid:

1. We are again going to change our perspective so we are looking at the right side of our box in a 2D *orthographic* view. I also find that adjusting small measurements like this is easier to do when adjusting the **Snap Grid** distance, which we can do using the menu shown in *Figure 16.12*. I am going to adjust the **Snap Grid** value to be half of my tolerance for a clearance fit, which is **0.5 mm**.

Figure 16.12: Adjusting the Snap Grid distance from a 2D orthographic view

2. Next, we are going to make the solid box shapes for the lid 1 mm smaller than the opening we cut when making the grooves. To do this, we need to make each side 0.5 mm smaller, as shown in *Figure 16.13*:

Figure 16.13: Creating wiggle room between the grooves and the lid

You could make these adjustments by entering the measurements, or by using the **Ruler** tool, but I find that the easiest method is to drag the transformation handles of the box using the distance set in **Snap Grid**. Because we set this to **0.5 mm**, each time I drag a handle, it moves by exactly this amount, as we initially learned in *Chapter 2*.

3. It can sometimes be difficult to see how the different parts fit together when designing multi-part projects like this. I like to set some of my shapes to be a **Transparent** color so that I can inspect how the different parts fit together, as shown in *Figure 16.14*:

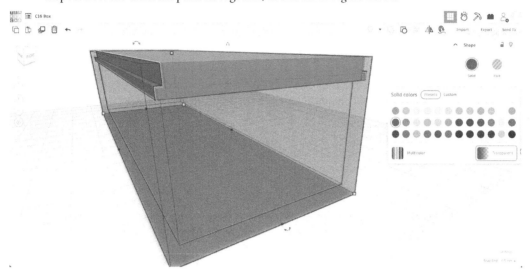

Figure 16.14: Creating transparent shapes and inspecting the model

4. Once we've ensured that the proper clearance has been created between the lid shapes and the grooves of the box, we can group the two boxes used for the lid to make a solid part, as shown in *Figure 16.15*:

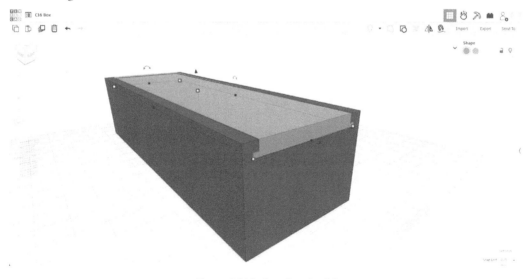

Figure 16.15: Creating the lid

At this stage, we could consider this project to be completed as we've successfully modeled a box and a lid. However, I think there's still room for improvement as we consider how this project could be made to be a bit more aesthetically pleasing and unique.

Adding artwork

Adding images and text to the lid of our box will enhance the overall appearance and could also be a useful feature to identify whose box it is, or what's inside it. As learned in *Chapters 5* and *8*, there are a lot of different ways that we can enhance our project using text and imported images.

Adding text features

The first thing we are going to look at are the text features that can be used to label and personalize the lid of our box:

1. As learned in *Chapter 5*, there are a few shapes that can be used to create text in our projects. For my box, I've chosen to use the **script** text shape, as shown in *Figure 16.16*:

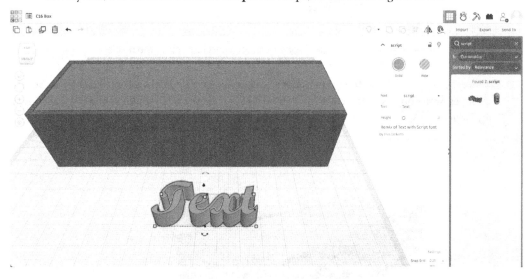

Figure 16.16: Adding the script text shape to our project

2. After adjusting the parameters for the text shape to label your box as desired, you can reposition the text shape so that it is on the surface of the lid, as shown in *Figure 16.17*:

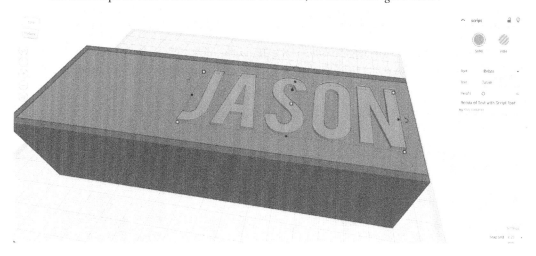

Figure 16.17: Adjusting and repositioning the text shape

In the last chapter, we looked at how text features could be *extrusions* or *pockets* when we designed the base for our trophy. The same concepts could be applied here as you decide how the text will be modeled. I've chosen to design my text feature so that it extrudes from the surface of the lid by 1.5 mm.

Once you've finished adjusting the text feature, we can move on to importing some images to further enhance our design.

Adding image features

As discussed in *Chapter 8*, we can find or create *SVG images* that can be imported into our Tinkercad designs. Once you've chosen an image, you can add it to your box through the following steps:

1. Click **Import** in the top-right corner of the design window to open the options for importing a file. You can then select the SVG image you would like to bring into your model, as shown in *Figure 16.18*:

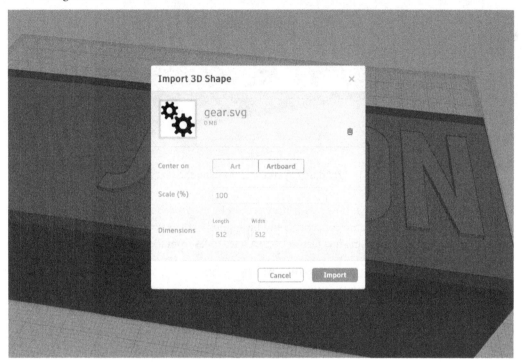

Figure 16.18: Options for importing SVG images

2. After pressing **Import**, you can reposition the imported image onto the surface of your lid. An effective way to do this is to use the **Cruise** tool to position the shape on the lid, as discussed in *Chapter 7*. You can then hold the *Shift* key on your keyboard while dragging the transformation handles to proportionally scale the image to fit on your lid, as shown in *Figure 16.19*:

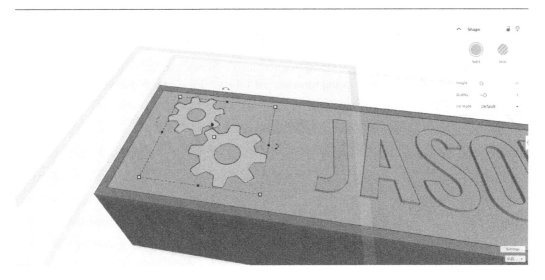

Figure 16.19: Adding an image to the lid

3. Lastly, we can group the image, text, and lid shape to create a solid and finished lid for this project, as shown in *Figure 16.20*:

Figure 16.20: Completing the lid

With the box and lid parts now complete, we are ready to export and manufacture this model using a 3D printer!

Exporting and manufacturing our models

Before we export our models for 3D printing, we need to reposition them to prepare them for manufacturing. To do this, I am going to move the lid out of the box and lower it to touch the **Workplane**, as shown in *Figure 16.21*:

Figure 16.21: Repositioning the parts for exporting

Next, we can press the **Export** button in the top-right corner of the design window to open the options for 3D printing our design, as shown in *Figure 16.22*:

Figure 16.22: Options for exporting our models

As discussed previously in *Chapter 14*, you can choose the best file format for your manufacturing resources, or choose to manufacture your project using a 3D printing production service instead. You also want to decide whether you are going to print the lid and box parts together, or whether you would like to export them as separate files for manufacturing with different materials like we did in the last chapter for the trophy parts.

I've chosen to download the lid and box together as a single **.STL** file. After downloading this file, I imported it into the CAM program for my 3D printer, as shown in *Figure 16.23*:

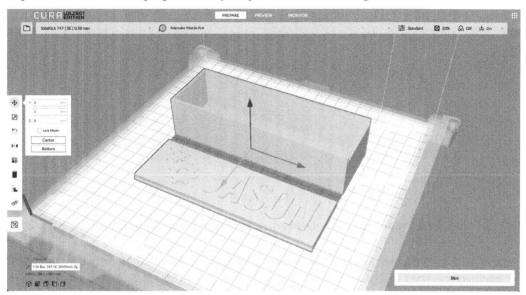

Figure 16.23: Viewing the exported STL in Cura LulzBot Edition

With the model loaded in my 3D printer's CAM software, we can adjust our print settings based on the needs of this project. As initially introduced in *Chapter 14*, there are many different settings we could adjust, but I will be highlighting a few important considerations for this box project:

- **Material**: As in the previous chapter, we need to ensure we've selected the correct material to properly prepare this model for manufacturing. I will be using PLA to print both parts of the box.
- **Quality**: Unlike the trophy made in the last chapter, I am going to print the box parts at a slightly lower resolution to save some time while still achieving a level of detail that is acceptable for this project.
- **Infill density**: For the box, I will be leaving the default infill density of 20% to ensure that the walls are strong enough to perform well, and to cut down on any unnecessary material.
- **Supports**: Because we designed our grooves to be small overhangs, no support material should be needed for these parts.

After adjusting your settings, we can slice our model and preview it for inspection, as shown in *Figure 16.24*:

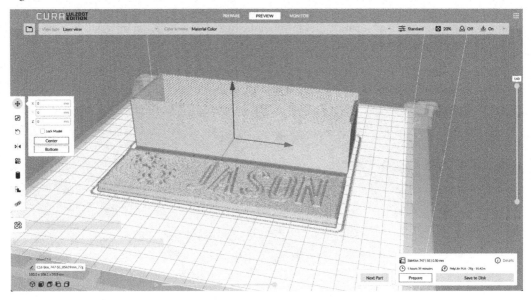

Figure 16.24: Previewing the sliced model to inspect it for manufacturing

Remember, previewing the model and the first layer is an important step to ensure that our prints will be manufactured successfully! Once you've inspected your project and double-checked all the print settings, you can save the *Gcode* file to your 3D printers to start manufacturing, as shown in *Figure 16.25*:

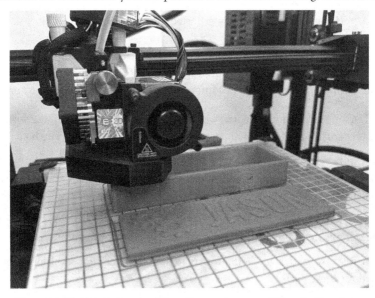

Figure 16.25: 3D printing the lid and box parts on an FFF-style 3D printer

After 3D printing has finished, there may be a few post-production steps needed based on how you manufactured your models. You may find that there are some excess filament strings in the grooves of the box if you are using an extrusion-like 3D printer like me, and these can carefully be cleared out with tweezers, a hobby knife, or even sandpaper.

There may also be times when the 3D-printed surface is not a desirable finish for your projects. Many professional model makers and DIY'ers enjoy sanding 3D-printed parts to be able to apply a smooth painted finish, as shown in *Figure 16.26*:

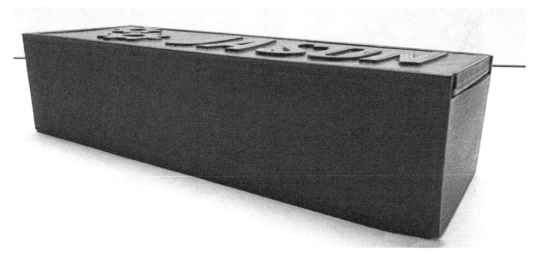

Figure 16.26: A 3D-printed part that has been sanded and painted

There are many different methods and strategies that can be employed when striving to paint your 3D models, but I often find that using a matte printing material, such as *Polymaker's PollyTerra* filament works better than glossy ones. You can also apply putty to the side of a sanded 3D print to smooth out the layers. Epoxy modeling putty typically works well, as does wood filler when diluted with some water - which is my preferred method. After sanding and smoothing the surface of your prints, you can apply a self-etching primer using a spray can to create a smooth surface prepped for painting, like the one shown in *Figure 16.26*.

After cleaning up the parts or refinishing the surface as desired, you can assemble the box and begin to use it for storage, as shown in *Figure 16.27*:

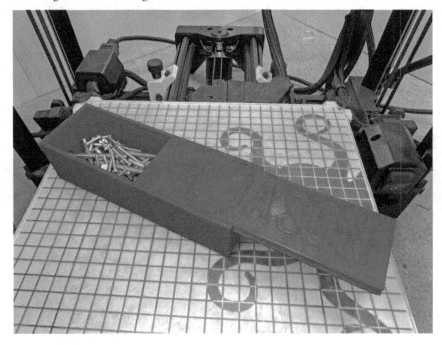

Figure 16.27: Testing and using the box with a sliding lid

And that just about wraps up this project! Hopefully you found that your measurements were accurate and that your box does well in storing the items that you've chosen. But if you ran into trouble, remember that making mistakes does not mean that you have failed.

We can learn from our mistakes as we continue to gain skills and abilities in designing real-world products and solutions like this multi-part box. If your items don't fit, or if the lid is a little too tight or too loose, that's OK! Head back into your Tinkercad design and make some adjustments to your model, and consider adjusting the print settings in your CAM program until you find greater success.

Remember that 3D printers are still predominately used as prototyping machines so that professional designers and engineers can make mistakes and test out their ideas quickly and cheaply. As we gain these skills, we can also implement the same practices.

Summary

This chapter was an exciting opportunity for you to again apply your skills and abilities in using CAD, CAM, and 3D printing technology to bring your designs into the physical world. I have always found useful projects, such as the box we created in this chapter, to be even more satisfying than the novelty items and trinkets that are so commonly created using hobby 3D printers. Hopefully you too have enjoyed this project and are motivated to continue with these challenges!

As we strive to create projects like the multi-part box with a sliding lid, we must take great care in planning out our projects by taking important measurements of the things our designs must interact with, as well as considering all steps of the manufacturing process throughout the design stages.

If we didn't accurately measure the items that were to be stored in our box, or accurately incorporate tolerances to suit the type of 3D printer and material we are using, then this might not be successful. The most successful designers are ones who understand each part of the production process, from conceptualizing the idea to bringing that idea to life so that it can perform as intended.

It's also important to remember that there are infinite ways to customize these types of projects, and you as the designer have the power to do so! This can be from using different shapes and adjusting scale, choosing different artwork or text features, creating your own shapes from scratch, as discussed in *Chapter 9*, or even sanding and applying a painted surface in the postproduction stages. The possibilities are truly limitless as we strive to design and fabricate our own unique creations.

We will continue along this path as we look at another project that relies on dimensions and tolerances to work, but this time with a twist! When you're ready, turn the page to get started.

17
Modeling an Ergonomic Threaded Jar

This chapter will allow us to continue to expand our skills through another practical application that requires a deeper understanding of tolerances within our design as we model a threaded jar! To further enhance our design skills, we will also consider the functionality of this product as we consider how to improve the **ergonomics** of our design through the following topics:

- Modeling the jar
- Modeling the lid
- Modeling the threads
- Adding ergonomic features
- Exporting and manufacturing the jar

In addition to considering tolerances between the different parts of our design as we did in the previous chapter, we will also be taking advantage of shape generators and advanced modeling techniques to create this real-world product. Also, like the previous project, we can design our threaded jars to be whatever size we want so that it is useful to store or transport things after manufacturing!

Technical requirements

The technical requirements will be the same as specified in the *Technical requirements* section of *Chapter 15*. However, you can find the example model shown throughout this chapter at this link: `https://www.tinkercad.com/things/gHKXG1PzIZ5-threaded-jar-model-from-chapter-17`

Modeling the jar

The first part of this project involves modeling the jar which will need to be scaled so that items will fit inside of it. Before you begin your CAD design, you should identify what you plan to store in your jar and take some preliminary measurements as we did in the previous chapter. This project will be scalable, which means you can always choose to increase or decrease the size of your jar later.

Once you've identified how big you would like this project to be, we can move into modeling the first part through the following steps:

1. After creating a new 3D design project in Tinkercad, choose a shape for your jar. It doesn't need to be cylindrical; I have chosen to start with a **Polygon** shape, as shown in *Figure 17.1*:

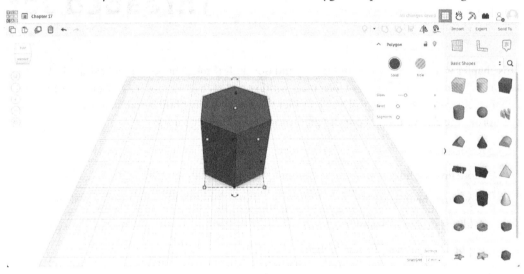

Figure 17.1: Choosing a shape to model the jar

2. Next, we can dimension this shape so that it will be big enough to store whatever items we have chosen. I have dimensioned the **Polygon** in my design to be 50 mm in length, width, and height, and also to have 8 **Sides** and a **Bevel** setting of 2 mm as shown in *Figure 17.2*:

Modeling the jar 299

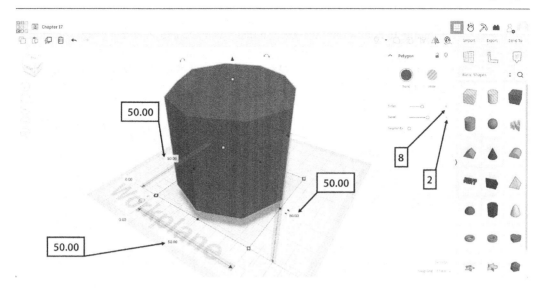

Figure 17.2: Dimensioning the polygon

3. The bevel will add a nice aesthetic to the jar, but we do not want to bevel to be at the top where the lid will be so that we can create a flush connection between the two parts. To fix this, we can add a **Box** that is set to be a **Hole**, and use it to cut the top of the **Polygon** by **grouping** the two shapes together as shown in *Figure 17.3*:

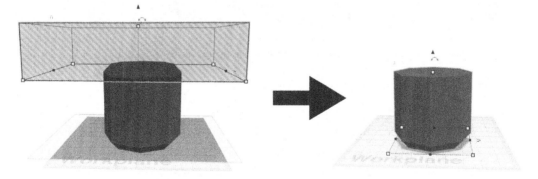

Figure 17.3: Cutting the top bevel off the Polygon

4. Next, we are going to hollow out this shape so that it will become an empty jar to store things in. There are a number of ways to do this, but the most efficient way to **shell** a unique shape is to make a smaller copy of the shape and use it to hollow out the inside. This can be done in the following steps:

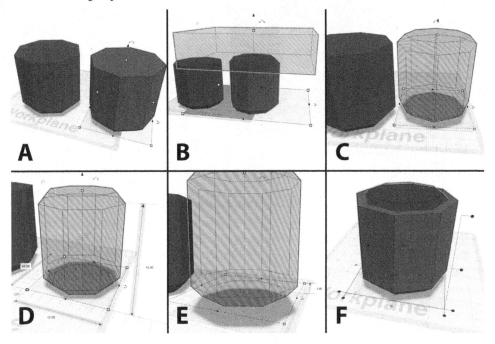

Figure 17.4: Hollowing out the polygon using another polygon

Figure 17.4 shows these steps:

- A. Make a copy of the polygon shape.
- B. Ungroup the copy and remove the hole box so that the top is once again beveled.
- C. Set the copy polygon to be a **Hole** shape.
- D. Scale the copy down so that it is 5 mm smaller in all dimensions.
- E. Raise the 5 mm smaller copy off the workplane.
- F. Align the copy shape with the original solid shape so they are centered in the *X* and *Y* axes.

G. **Group** the two shapes together to create a hollow shape, as shown in *Figure 17.5*:

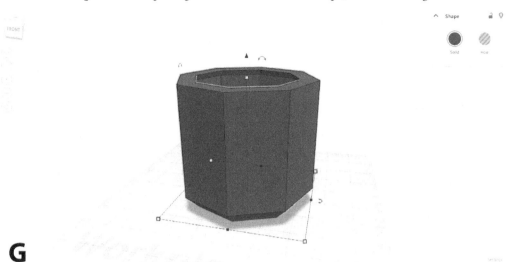

Figure 17.5: A hollowed polygon shape

When working with hollow shapes like this, it can be difficult to inspect the designs for any possible flaws as we cannot see what's happening inside the outer shell. A good strategy to use is to set the **Solid** shape to be a **Transparent** color so that we can see inside of it, as shown in *Figure 17.6*:

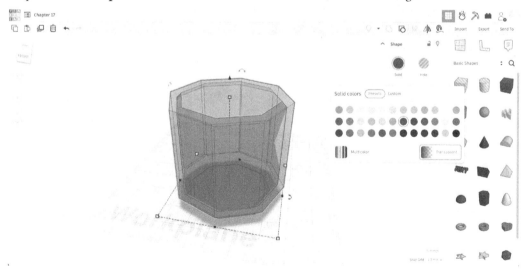

Figure 17.6: Setting the shape to be transparent so that it can be inspected

Figure 17.6 allows us to see the inner pocket created earlier in *Step 4*. We can also see that the top bevel left on the inner pocket has created a **fillet** along the top edge. This will allow for the pocket to be 3D printed without needing **supports** as the **overhang** was reduced, which we learned about earlier in *Chapter 11*.

Now that we've completed the jar portion of this project, we can move on to creating the lid next.

Modeling the lid

Like the jar part of this project, we can model the lid using just about any shape we would like. While the lid and the jar do not need to be symmetrical to one another, I find that using the same shape for both parts can make for an aesthetically pleasing result. As such, I will be creating the lid based on the eight-sided polygon used for the jar created previously:

1. The first step for modeling the lid is to grab a shape to create the lid with. I have chosen to copy the jar to use as a starting point for my lid, as shown in *Figure 17.7*:

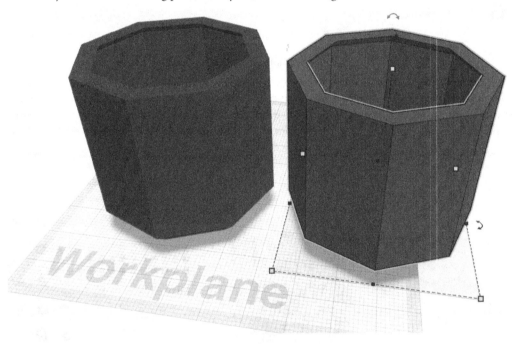

Figure 17.7: Copying the jar to start creating the lid

2. Next, we can flatten the lid polygon to be less tall, though the lid can be as tall as you would like as this is not a critical dimension. I have chosen to dimension the height of my lid part to be 15 mm, as shown in *Figure 17.8*:

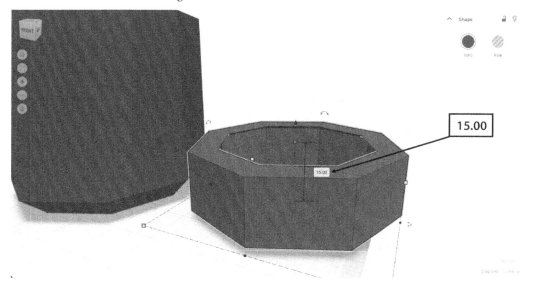

Figure 17.8: Dimensioning the height of the lid

3. Lastly, we do not need the lid to be hollow like the jar. **Ungroup** the lid shape so that it can be separated from the hole polygon shape, then delete the hole polygon as it is not needed for the lid part, as shown in *Figure 17.9*:

Figure 17.9: Removing the hole shape from the lid

With a simple polygon created to match the jar shape we made earlier, the lid is ready for threads, which will be made in the next steps.

Modeling the threads

While we won't find a helix or thread-like shape in Tinkercad's basic shapes, making threads and screws is still possible in Tinkercad. One way to do this is to source a threaded STL such as a bolt that can be imported into our designs, as we discussed earlier in *Chapter 2*. However, importing shapes is not the only way to make this custom feature. We can also take advantage of a **Shape Generator** as we initially discussed in *Chapter 9* to make a threaded union between the lid part and jar part we created earlier by following these steps:

1. First, we can type `thread` in the search feature in the shapes library to find the **ISO metric thread** shape, then bring this Shape Generator into our design as shown in *Figure 17.10*:

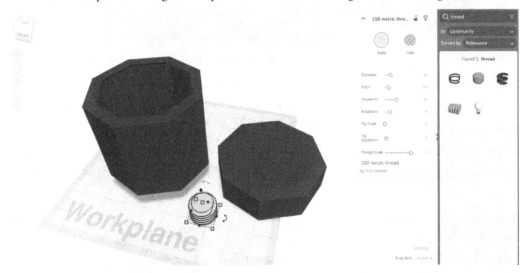

Figure 17.10: Searching for the thread Shape Generator

2. We can then adjust the size of the **ISO metric thread** shape so that its diameter fits within our jar. This should match the size of the shape we used to hollow out the jar, which was 45 mm in diameter for my polygon model. We also can adjust the **Segments** count, which adjusts how smooth the shape is, as well as adjust the number of **Rotations**, which controls the thread count. I have adjusted these parameters to be 32 and 3 respectively, as shown in *Figure 17.11*:

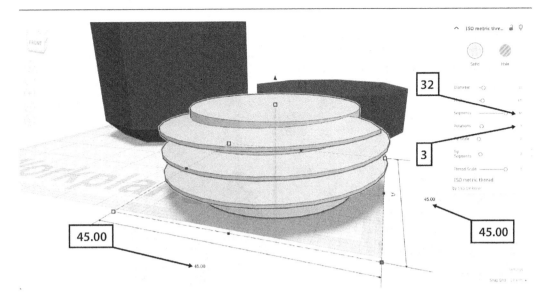

Figure 17.11: Adjusting the parameters for the thread shape

3. Next, we can move and align the **ISO metric thread** shape so that it is centered on top of our lid model, as shown in *Figure 17.12*:

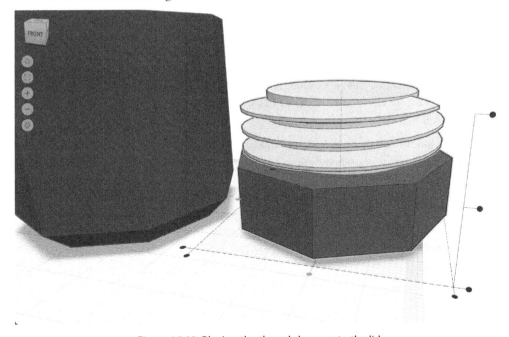

Figure 17.12: Placing the thread shape onto the lid

4. We are going to use the **ISO metric thread** shape to create both the male and female threads for the lid and the jar. To do this, we are going to make a copy of the thread shape using the **Duplicate and repeat** tool, then set the copy to be a **Hole**, and lock the shape in place so it cannot be edited, as shown in *Figure 17.13*:

Figure 17.13: Making a locked copy of the thread shape

The reason we are making a copy of the shape like this is to allow us to easily adjust the **tolerances** between the male and female thread shapes, similar to how we adjusted the grooves for the slide box in the previous chapter. You can lock a shape in place by clicking on the image of the lock in the shape parameters, or by pressing *Ctrl + L* on your keyboard. Once a shape is locked, a thin purple shadow will appear lightly around the locked shape, as shown in the image to the right of *Figure 17.13*, and the locked shape now cannot be edited until it is unlocked in the same way it was locked.

5. With the hole thread shape locked, we can adjust the solid thread shape to be the male threads that will fit within, as shown in *Figure 17.14*. In the shape parameters for the **Solid** threads, we can adjust the **Thread Scale** value to 0.5, which will add some wiggle room between the solid threads and the hole threads. We can then make the threads 1 mm smaller for all sides in both the *X* and *Y* dimensions or 2 mm smaller across the overall width and height, which will create a **clearance fit** (as discussed previously in *Chapter 12*) between the solid thread shape and the locked hole thread shape, as shown in *Figure 17.14*:

Modeling the threads 307

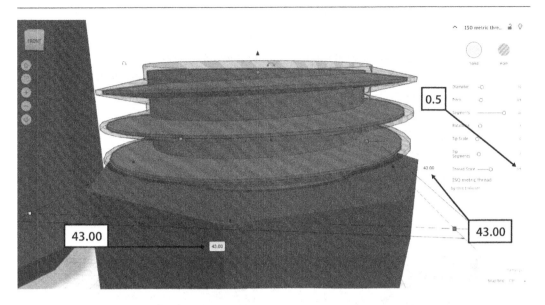

Figure 17.14: Creating a clearance fit between the thread shapes

6. Next, we can unlock the hole thread shape and move it into the jar shape, as shown in *Figure 17.15*:

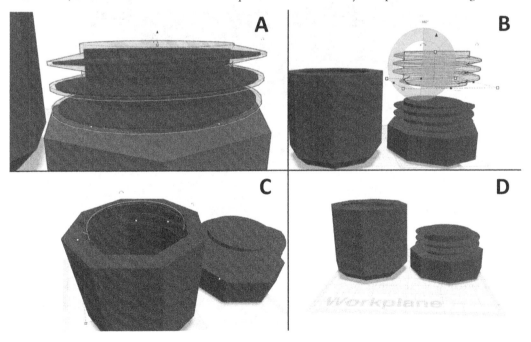

Figure 17.15: Adding threads to the jar part

Figure 17.15 shows these steps:

A. Unlock the hole thread shape so it can be edited by selecting the locked shape, then clicking on the lock icon in the shape parameters, or pressing *Ctrl + L* on your keyboard.
B. Raise the hole thread shape off the lid shape, then rotate the hole thread shape 180 degrees.
C. Move and align the hole thread shape so it is centered in the *X* and *Y* axes on top of the jar part. The threads should start flush with the top of the jar, as shown in *Step C* of *Figure 17.15*.
D. **Group** the hole thread shape with the jar part to cut the female threads.

As discussed previously, it's always a good idea to do a test fit between your parts to make sure there aren't any issues before moving to the next step. To do this, set the jar part to be **Transparent** and move the lid into position to inspect how the parts fit together, as shown in *Figure 17.16*:

Figure 17.16: Performing a test fit of the jar and lid parts

The threads should fit within each other with tolerance between them, as seen in *Figure 17.16*. It is also important that the bottom of the lid is touching the top of the jar when the threads fit together, or else the jar won't fully close or tighten. You can ungroup the hole threads from the jar part and adjust the vertical position until the fit is just right.

Once the threads look as they should, we can move the lid back off to the side before moving into the last part of the design stages for this project.

Adding ergonomic features

Ergonomics typically describes the study of people's efficiency in a working environment, but in the world of design, we often consider ergonomics to be features that make a product easier or more efficient for humans to use. If you've ever held a coffee mug with a rubberized sleeve, carried a bag with a padded handle, or used a tool that was knurled for better grip, you've experienced ergonomic designs before. Let's now apply these concepts in our design to improve the functionality of our jar, just like the professionals, through the following steps.

Improving the jar part

Starting with the jar, let's add a raised pattern to make this part easier to hold onto:

1. First, select the **Half Sphere** shape and bring it into your design. We can scale this shape down so it fits onto the side of our jar, and use the **Cruise** tool to position the shape so the rounded part is facing outwards off one side of the jar, as shown in *Figure 17.17*:

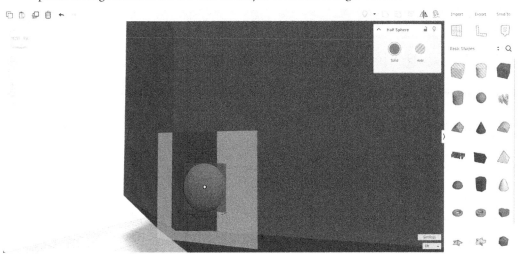

Figure 17.17: Adding a half sphere to one side of the jar part

2. Using the **Duplicate and repeat** tool, make copies of the **Half Sphere** and pattern the copies across the side of the jar in a row, as shown in *Figure 17.18*:

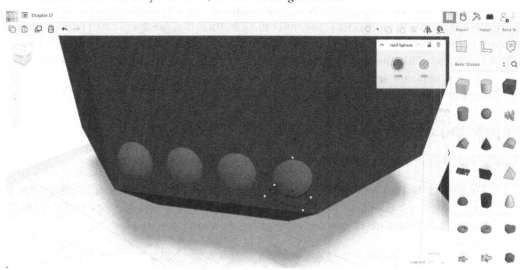

Figure 17.18: Patterning a row of half spheres on the side of the jar

3. Select all the half spheres in the row, then use the **Duplicate and repeat** tool to add rows so the entire side of the jar is covered, as shown in *Figure 17.19*:

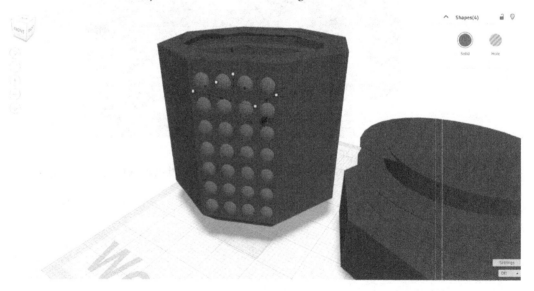

Figure 17.19: Completing the pattern of half spheres on the side of the jar

4. Next, select all the half spheres and group them together so that they are easier to select and move. Once grouped, use the **Duplicate and repeat** tool to copy the group of half spheres, then rotate and move the copy so the shapes are aligned with the next side of the jar, as shown in *Figure 17.20*:

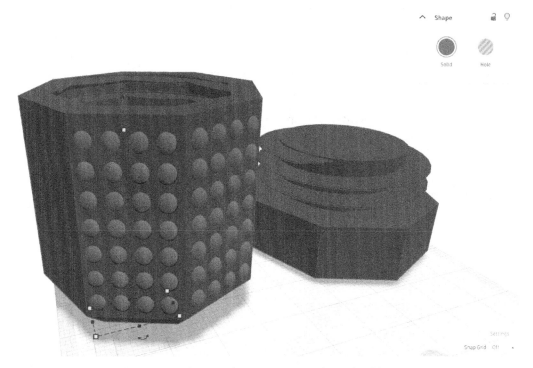

Figure 17.20: Copying the pattern to another side of the jar

5. Use the **Duplicate and repeat** tool to continue this pattern of shapes around the entire jar, then **Group** the half spheres with the body of the jar to complete this part, as shown in *Figure 17.21*:

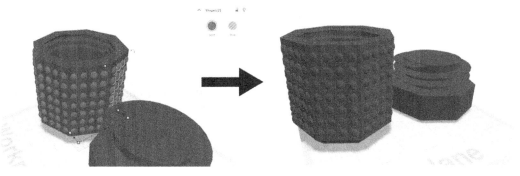

Figure 17.21: Finishing the pattern and grouping all shapes with the jar

The half spheres will add a bumpy texture around the jar, which not only may increase its aesthetics, but also make the jar easier to hold onto, adding to its ergonomic functionality too. We can create another pattern on the lid for a similar result on the other part of this design.

Improving the lid part

Like the jar part, we are going to create a pattern texture around the lid so it is easier to grip and turn as we open and close the jar. Instead of creating the same dotted pattern we used on the jar, we can create notches around the lid that will have a similar ergonomic effect but also add to the overall aesthetics of the design:

1. Start by bringing the **Round Roof** shape into your design. We then need to set this shape to be a **Hole** shape, rotate it 90 degrees so it is vertical, and scale it so it is tall and narrow, as shown in *Figure 17.22*:

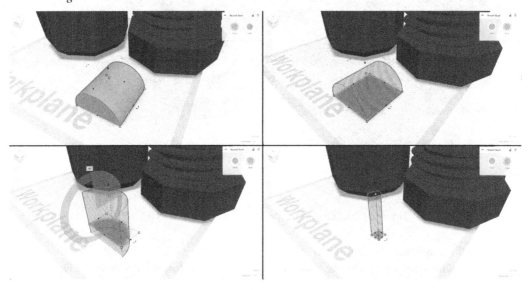

Figure 17.22: Adding and adjusting the round roof shape

2. Next, change your perspective so you are looking down on the lid from the **Top** view, and adjust your view so you are looking at the part in a flat 2D *orthographic* view, as discussed in *Chapter 3*. This will make it easier to move the **Round Roof** shape onto the edge of one side of the lid shape, as shown in *Figure 17.23*:

Adding ergonomic features 313

Figure 17.23: Moving the round roof onto one side of the lid

3. Next, use the **Duplicate and repeat** tool to copy and move the **Round Roof** so that an entire side of the lid is patterned, as shown in *Figure 17.24*:

Figure 17.24: Patterning the round roof shape on one side of the lid

4. Like we did when creating the jar part, we can then use the **Duplicate and Repeat** tool to pattern groups of the round roof shapes on each side of the lid, then group these shapes with the lid body to cut notches into the lid, as shown in *Figure 17.25*:

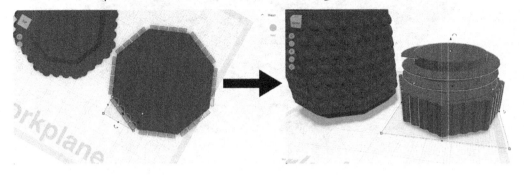

Figure 17.25: Completing and grouping the round roof patterns with the lid part

With both the ergonomics and aesthetics of the jar and lid parts now improved, it is time to move into exporting and manufacturing this model so we can put it to use!

Exporting and manufacturing the jar

As we've done in the past two practical applications (in *Chapters 15* and *16*), the first step to manufacturing this project is to export or share the design depending on how we plan to manufacture it. After deciding on whether you want to manufacture the jar and lid parts together, you can select your shapes and press **Export** in the top-right corner of the design window to open the options for exporting, as shown in *Figure 17.26*:

Figure 17.26: Options for exporting the design

If you plan to 3D print this project yourself, you can choose a file format to download the model based on your available resources, as discussed earlier in *Chapter 14*. Alternatively, you could use a 3D printing service to manufacture your design under the **3D Print** options of the window shown, also discussed previously in *Chapter 14*.

If you're manufacturing this project yourself, the next step would be to bring the files into the **slicer** program for your 3D printer like me, as shown in *Figure 17.27*:

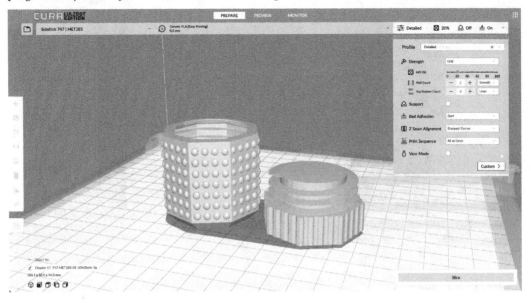

Figure 17.27: Loading the design files in Cura LulzBot Edition

In *Figure 17.27*, we can see that I imported the jar and lid files into *Cura LulzBot Edition* for my 3D printer. As discussed in previous chapters, we need to adjust the print settings based on the needs of our project and our available resources. Some things to consider for this project are as follows:

- **Material**: As we've discussed before, we need to choose a printing material that suits the needs of this project. If you're storing common items in this jar like me, using PLA would be a suitable choice. Alternatively, you may want to choose a temperature or UV-resistant material if you plan to keep your jar in your car, or perhaps a food-grade material if you're holding snacks instead.

- **Quality**: Depending on the size of your jar, you may need to select a medium-high resolution to retain the detail in the ergonomic features of the design, as well as to print the threaded parts successfully.

- **Infill density**: As this jar has several small details, we don't want to lower the infill density too much. An infill density of 20% should allow for the jar to be rigid enough to function well without wasting unnecessary material.

- **Supports**: Because we utilized bevels and because threads are naturally building, no supports should be needed to print this project successfully at a higher print quality.

With your settings adjusted, you can slice and preview your model to check for any possible issues. Always remember to check the first layer of your print, as shown in *Figure 17.28*:

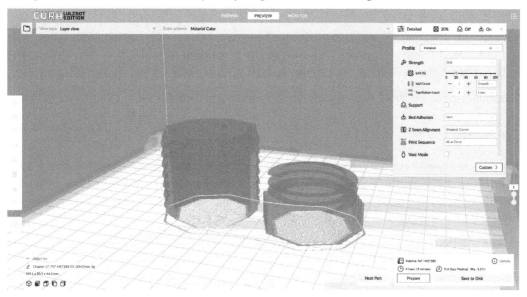

Figure 17.28: Inspecting the first layer of the print preview

Once all settings have been checked, you can save the *Gcode* file to your printer and manufacture this project! Once completed, you should have an easy-to-use multi-part project to store items in, as shown in *Figure 17.29*:

Figure 17.29: Two completed ergonomic threaded jars

As we didn't need any supports for this project, there shouldn't be any post-production steps needed to prepare this model before use. Test the functionality of the jar by threading the lid onto the jar, and check to see whether the jar successfully stores the items you chose earlier in this chapter.

If the threads don't quite fit, adjust the tolerance between the threads in your design and attempt to reprint the model until you find success. You can also proportionally scale this project to make the jar bigger or smaller, and a larger jar may be easier to 3D print successfully.

Remember, making mistakes is a crucial part of the learning process!

Summary

This chapter allowed us to create another useful project while also providing opportunities to gain greater proficiency in using some of Tinkercad's more advanced tools and features. Incorporating measurements and tolerances in our design was an important step to finding success in this application and one that we will continue to practice in the coming chapters.

We also put the **Duplicate and repeat** tool to good use in this chapter as we made patterns in our design, as well as to copy our shapes more efficiently. The unique angles on the parts of our model also allowed us to use the **Cruise** tool to efficiently move shapes onto different workplanes in the design more easily than if we tried to manually rotate and position the parts without it.

This chapter also allowed us to expand our understanding of real-world design principles as we not only considered the aesthetics of our project but the ergonomic functionality of the parts too. These concepts are not always needed to find success in a project, but they can always be used to improve the overall functionality and performance of the things we intend to use.

When you're ready to expand on these concepts through a project that requires a bit more creative thinking, turn the page to make our biggest project yet as we design, fabricate, and play a multi-part puzzle!

18
Building and Playing a 3D Puzzle

In this chapter, we will once again be putting our skills to the test as we design and build a 3D puzzle! This practical application will not only allow us to strengthen our skills in using Tinkercad to design models for 3D printing, but also look at some tools and methods for effective design which have not yet been discussed. These challenges will be broken down into the following topics:

- Making the pieces
- Making the joint template
- Adding the joints
- Incorporating artwork
- Preparing, exporting, and manufacturing the puzzle

As we use our design and modeling skills, we must also consider how this project can be designed to make it challenging for the user to play so that it may be an effective product. This will allow us to incorporate real-world considerations for product design, as well as continue to expand our abilities in successful modeling and manufacturing techniques.

Technical requirements

Later in this chapter, we will be incorporating artwork and images into our designs, as discussed in *Chapter 8*. You can use programs such as *Google Drawings* or *Vectr* to make your own images that can be imported, or you can browse image repositories such as *SVG Repo* to find ones to use in your design. An editable model of the puzzle shown throughout this chapter can be accessed on Tinkercad at `https://www.tinkercad.com/things/2OXWUA6tEoX-puzzle-model-from-chapter-18`.

Making the pieces

There are many kinds of 3D puzzles out there, and by using the skills you've gained in this book you can strive to make a unique creation of your own! In this chapter, we will be designing a 3D puzzle that has five pieces, and one that is designed to not only be engaging to use but also well suited for manufacturing using 3D printing techniques. To get started, we need to create the pieces for the puzzle through the following steps:

1. We are first going to start by adding a **Box** shape to our design that is the overall size of our puzzle. You can make this puzzle as big or as small as you like, but I've chosen to model mine to be 100 mm in length, width, and height, as shown in *Figure 18.1*:

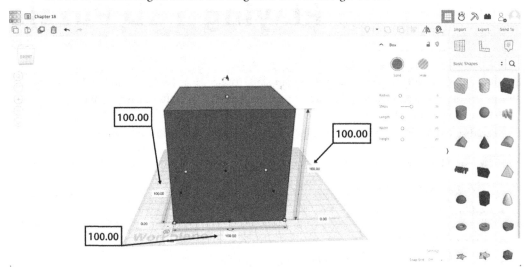

Figure 18.1: Starting the puzzle with a Box shape

2. This puzzle is also going to have a piece that divides the puzzle down the middle, and this piece can be a unique shape based on your own preference. I've chosen to use a **Polygon** shape with 6 sides and the dimensions of 30 x 30 x 100 mm in length, width, and height, as shown in *Figure 18.2*:

Making the pieces   321

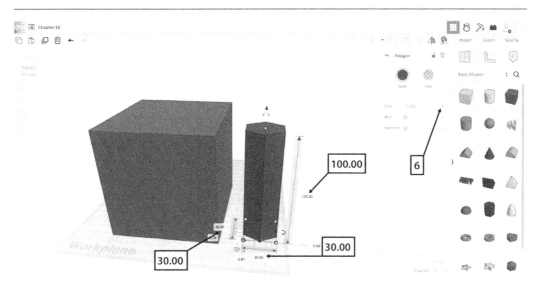

Figure 18.2: Adding a second shape for the puzzle pieces

3. As we learned in *Chapter 12*, we need to add **tolerances** to multi-part designs so that they will fit together after manufacturing them with 3D printers. To do this, we are going to first rotate the **Polygon** shape 90 degrees, then make a copy of the shape and dimension it to be 1 mm bigger in length and width, as well as set it to be a **Hole** shape. We can then move the **Hole** shape so that it is centered in the box shape we created earlier using the **Align** tool, as shown in *Figure 18.3*:

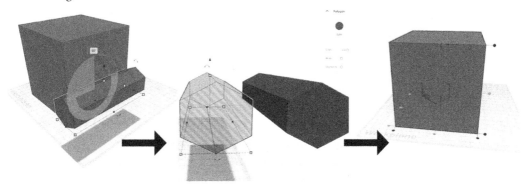

Figure 18.3: Making a Hole shape with tolerances added

To create the other four pieces, we need to divide the **Box** shape into quadrants, which can be done through the next steps.

4. Create a **Hole Box** shape that is half the width and height of the original **Box** shape, and aligned with one corner of the box, as shown in *Figure 18.4*. To assist in aligning these shapes, remember that you can click the shape you want to align to after selecting the **Align** tool, as discussed in *Chapter 7*.

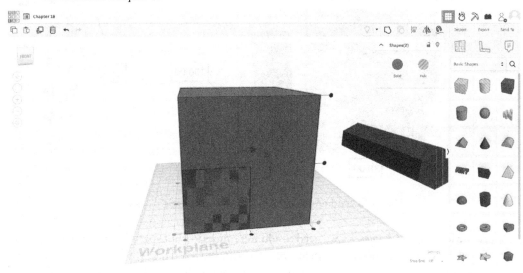

Figure 18.4: Creating the first Hole Box shape

5. Copy and move the **Hole Box** shape three times so that there are four total shapes aligned with each corner of the original **Box** shape, as shown in *Figure 18.5*:

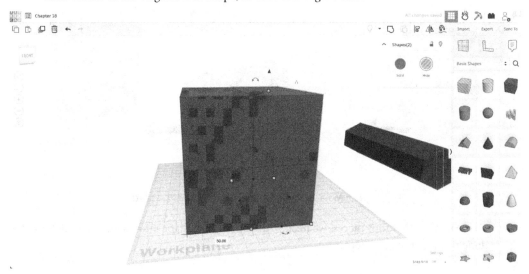

Figure 18.5: Making copies of the Hole Box shape

6. Duplicate the entire set of shapes, including the original **Box** shape, the four **Hole Box** shapes, and the **Hole Polygon** shape, so that there are four copy sets, as shown in *Figure 18.6*:

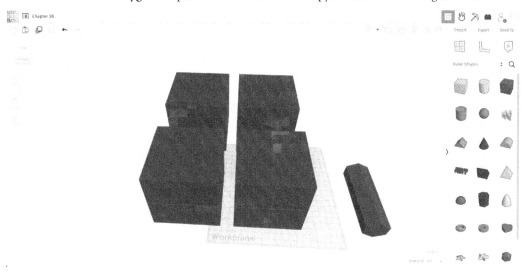

Figure 18.6: Making copies of the shapes

7. Then, delete a different **Hole Box** shape from each copy so that a different corner is showing across each set of shapes, as shown in *Figure 18.7*:

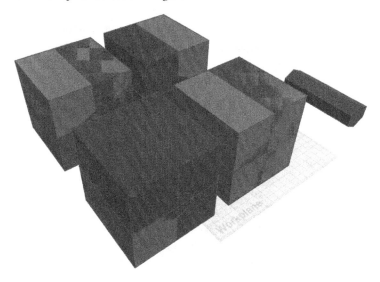

Figure 18.7: Preparing the shapes to cut the pieces

8. Lastly, group each set of shapes together one at a time to cut the four puzzle pieces, as shown in *Figure 18.8*:

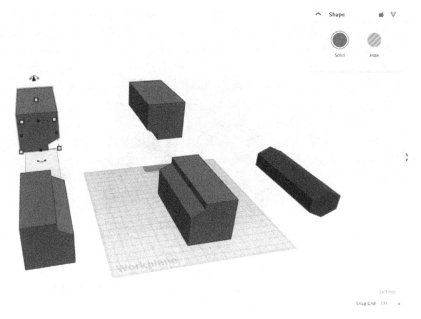

Figure 18.8: Creating the puzzle pieces

After creating all of the pieces, we can mock assemble the puzzle by moving the corner pieces together so they align around the original **Polygon** piece, as shown in *Figure 18.9*:

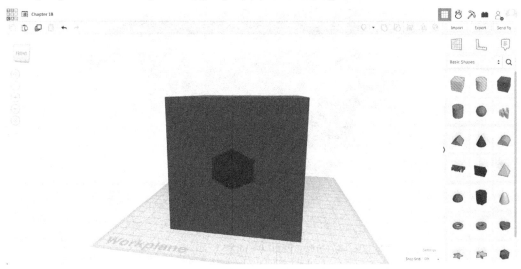

Figure 18.9: Aligning the puzzle pieces together

Next, it's time to create joints that will allow these pieces to interlock, but only in the correct order!

Making the joint template

For the pieces to be able to lock together, there needs to be some type of tab or extrusion that fits between adjoining pieces. This joint should also only allow for the pieces to fit together in one way so that the puzzle is more challenging to assemble. To create these joints, we are first going to create a template that will be copied and reused to make the connection joints on each piece later in the *Adding the joints* section:

1. To start creating the joints, we can first add a **Sphere** shape to our design with the default dimensions of 20 mm in LxWxH, as shown in *Figure 18.10*:

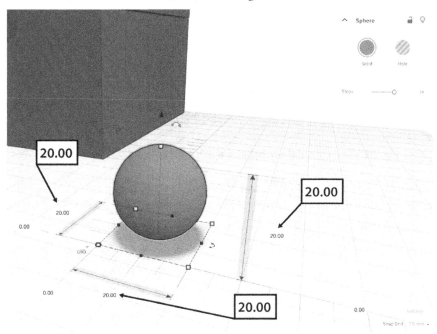

Figure 18.10: Adding a Sphere shape

We are using a **Sphere** shape to create the connection joints because its rounded structure will allow for the joints to be 3D printed without needing any support later on.

2. Next, we can copy the **Sphere** shape to create a **Hole Sphere** shape that is 1 mm larger in each dimension and aligned so the two **Sphere** shapes are centered within one another, as shown in *Figure 18.11*. This will allow us to create a pocket for each joint that has a 1 mm tolerance added, similar to what we did with the **Polygon** shape earlier in the *Making the pieces* section.

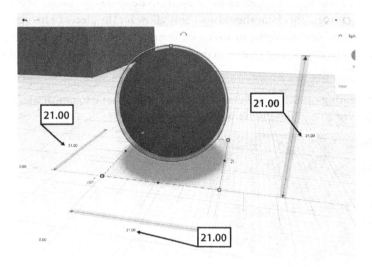

Figure 18.11: Creating a Hole Sphere shape over the solid Sphere shape

3. Next, we can copy the solid **Sphere** and **Hole Sphere** shapes, then move the copied shapes so they are 50 mm apart, as shown in *Figure 18.12*, so that each piece will have two joints:

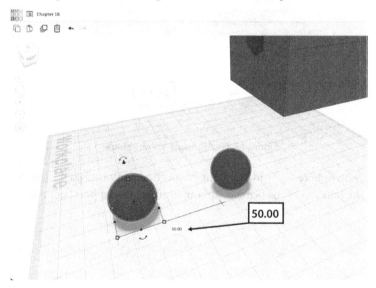

Figure 18.12: Creating a second set of Sphere shapes

4. We can then add a **Hole Box** shape that is dimensioned to be 100 mm in length, 0.5 mm in width, and 35 mm in height, and centered with the **Sphere** and **Hole** shapes, as shown in *Figure 18.13*, to complete the joint template:

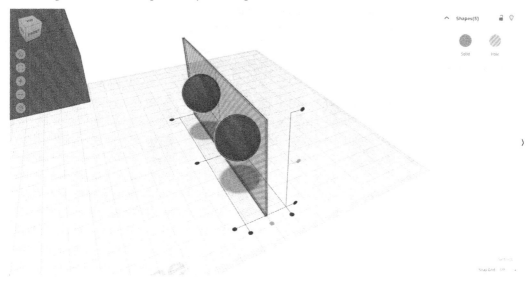

Figure 18.13: Completing the joint template with a Box shape

After completing the joint template, we can now easily access an original copy of the joint template shapes to make sphere-shaped joints between the different pieces of our puzzle easily in the next section.

Adding the joints

By creating and using a template to make the joints, we can be confident that the joint shapes will align from piece to piece while also saving time by not needing to recreate each solid and hole sphere shape needed for every joint. Using the template made in the last section, we can now create interlocking joints on our puzzle pieces through the following steps:

1. First, separate the puzzle pieces, as shown in *Figure 18.14*. We can also move the center **Polygon** piece off to the side as we will not be needing it for the next few steps.

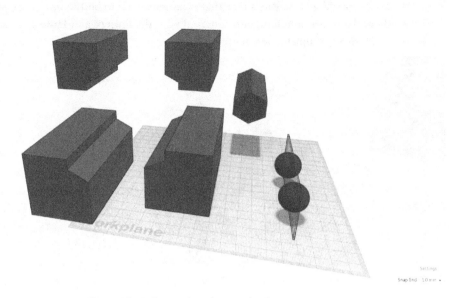

Figure 18.14: Separating the puzzle pieces

2. Make a copy of the shapes used in the joint template we made in the last section and move them so the shapes align with the face of a piece that connects to another piece, as shown in *Figure 18.15*. We can use the **Hole Box** shape included in the joint template to align with the face of the puzzle piece as this will create a tolerance gap between each piece later.

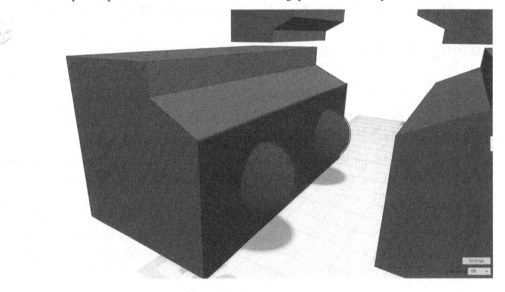

Figure 18.15: Aligning the template shapes onto the first puzzle piece

3. Using another copy of the template pieces, rotate and align the joint template pieces so that they align with the other connecting face of the puzzle piece, as shown in *Figure 18.16*:

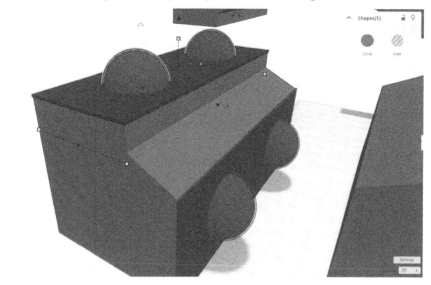

Figure 18.16: Aligning the second set of joints to the first puzzle piece

4. We then need to repeat *steps 2* and *3* for each puzzle piece so that all connecting faces include a joint template. If your pieces are symmetrical, you can copy the first piece prepared in *steps 2* and *3* for the other pieces of your puzzle. If your center shape is not symmetrical, you may need to copy and align the joint template for each piece manually so that all pieces are prepared with the template, as shown in *Figure 18.17*:

Figure 18.17: Placing the joint template on all four puzzle pieces

5. To make this puzzle work, a **Solid Sphere** joint needs to align with a **Hole Sphere** joint across two connecting faces. Additionally, each piece should connect using a different combination of **Hole** and **Solid** joints so that the puzzle must be solved in a certain way. To do this, work through deleting **Hole** and **Solid Sphere** shapes from the joint templates so that the connecting faces of each puzzle piece will align and interlock with the one next to it, as shown in *Figure 18.18*:

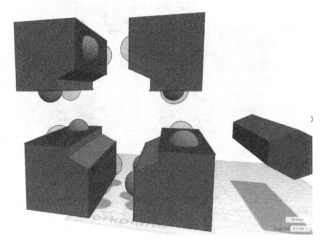

Figure 18.18: Using the template to make interlocking joints

6. Lastly, individually group each set of joint shapes with each puzzle piece to create the joints on the connecting faces, as well as to shave 0.5 mm off each connecting face to create tolerances between the puzzle pieces, as shown in *Figure 18.19*:

Figure 18.19: Grouping the joint templates with the puzzle pieces

After grouping the puzzle pieces individually with the joint pieces, we can bring the pieces back together to test the fit of the joints. This can be done more effectively by making the pieces **Transparent**, as shown in *Figure 18.20*:

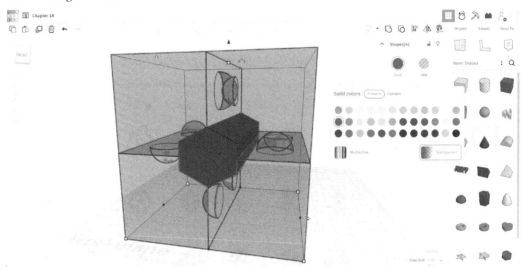

Figure 18.20: Test fitting the puzzle pieces and connecting joints

After inspecting each joint to ensure that they align from piece to piece, we can move on to complete this puzzle. To make it more complex, we now can add designs to the outside of the pieces that must align from piece to piece, like how the joints we've created also must align correctly to solve the puzzle!

Adding artwork

To further enhance this puzzle project, we are now going to add designs around the outside of the puzzle that will make it more challenging to solve. To do this, you need to decide what images, patterns, or themes you want to incorporate into your puzzle design, as well as decide whether this can be done using the shapes in Tinkercad or whether you need to look elsewhere.

As initially introduced in *Chapter 8*, we can import **SVG** images into our Tinkercad designs that are created in graphic programs, such as *Google Drawings*, or obtained from sites such as *SVG Repo*, as shown in *Figure 18.21*:

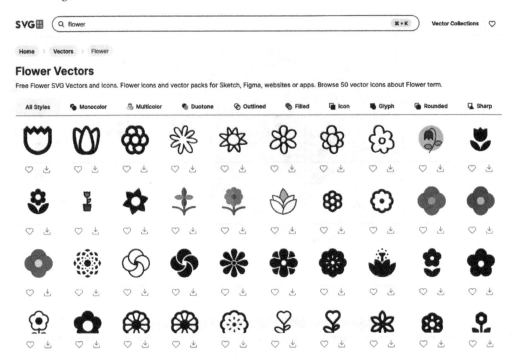

Figure 18.21: Browsing SVG Repo for vector images (image credit: www.svgrepo.com)

We can choose one image or pattern to include in our puzzle or choose a unique design for each side. Once we've obtained our images, we can add them to your puzzle through the following steps:

1. To import SVG images into our Tinkercad design, we need to press the **Import** button in the top-right corner of the design window, then select the SVG image we want to import, as shown in *Figure 18.22*:

Adding artwork | 333

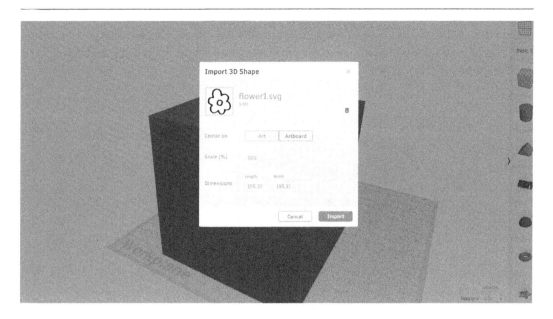

Figure 18.22: Options for importing an SVG image

2. To foster greater success during the 3D printing stages of this project, we want to ensure that these images just engrave the surface to prevent creating overhangs. To do this, we can move and scale the image so it fits onto a side of the puzzle, and set image to be a **Hole** with a height of 1 mm, as shown in *Figure 18.23*:

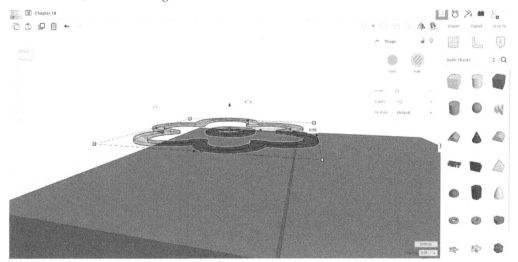

Figure 18.23: Moving and scaling the SVG image onto the side of the puzzle

3. We can then copy and scale the SVG image to make a patterned design on the side of the puzzle. Ensure that the images span across two puzzle pieces so that the design can only be completed if the puzzle is assembled correctly, as shown in *Figure 18.24*:

Figure 18.24: Creating a pattern on the side of the puzzle using images

4. Next, select all images used in the pattern and use the **Duplicate and repeat** tool to make a copy of these images in place, as shown in *Figure 18.25*:

Figure 18.25: Copying the patterned images

5. Individually select one set of images and group them with one of the connecting puzzle pieces, then group the other set of images with the other puzzle piece. This should engrave the images onto both puzzle pieces without grouping the two puzzle pieces together, as shown in *Figure 18.26*:

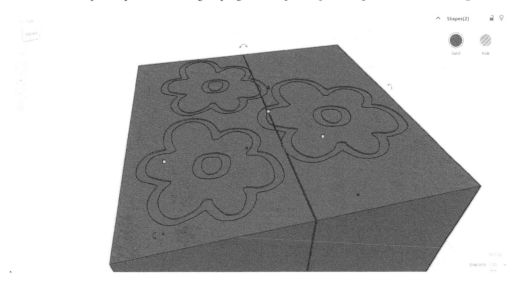

Figure 18.26: Grouping the images with the pieces

We can then repeat these steps using the same or different images across all four flat sides of the puzzle so that each side and set of pieces is unique, as well as adjust the colors of the pieces to simulate how the puzzle will look when manufactured, as shown in *Figure 18.27*:

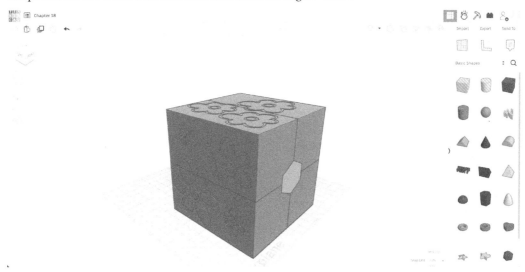

Figure 18.27: Patterned images across the sides of the puzzle

With the combination of connection joints and patterned images, there should now only be one possible way to solve this puzzle! It's now time to prepare and export our designs so that we can manufacture this project and test it out.

Preparing, exporting, and manufacturing the puzzle

As we've done in previous chapters, we must arrange the parts in our design so they can be 3D printed in the most optimal way. For the puzzle, this involves separating the pieces so the joints are facing out, and rotating the pieces so flat sides are touching the workplane, as shown in *Figure 18.28*:

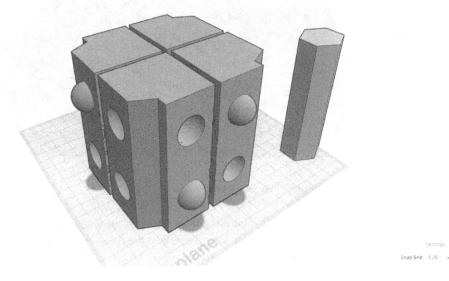

Figure 18.28: Arranging the puzzle pieces for 3D printing

After rearranging the pieces, we can proceed with exporting them for production. You need to decide whether you would like to 3D print the pieces together or separately in different colors as I will be doing. Select all the pieces you want to print together, then press **Export** in the top-right corner of the design window to open options for exporting the design, as shown in *Figure 18.29*:

Figure 18.29: Options for exporting the puzzle pieces

If you plan to 3D print this project yourself, you can choose a file format to download the selected shapes based on your resources, as initially introduced in *Chapter 14*. Alternatively, you could use a 3D printing service to manufacture your design, as discussed in the previous chapters as well.

If you're manufacturing this project yourself, the next step is to import the pieces of your puzzle into the **CAM** program for your 3D printer, as I did for mine in *Figure 18.30*:

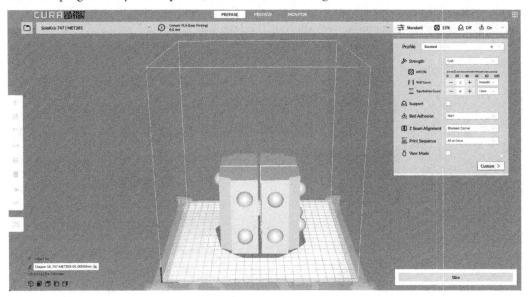

Figure 18.30: Loading parts of the puzzle in Cura LulzBot Edition

In *Figure 18.30*, you can see that I've imported four of the five pieces for my puzzle in *Cura LulzBot Edition* for my 3D printer as I plan to print the fifth piece separately in a different color on a single extruder 3D printer. As we've done in the previous practical applications of this book, we now need to adjust the print settings to suit the needs of this project. Some considerations for the puzzle project are as follows:

- **Material** – Simple novelty projects like this can be made effectively using easy-to-print rigid materials such as PLA
- **Quality** – A standard print quality should do well for this puzzle unless you've chosen to make it smaller than 50 mm in width; then, you may need to use a higher resolution so that the joints and images are printed successfully
- **Infill density** – We can lower the print density to 10% or 15% without risking failure for these puzzle pieces as they are very boxy shapes without any complex features
- **Supports** – Because we used spheres to create the joints, all overhangs will build upon themselves, removing the need for any support material

After adjusting your print settings for these parts, you can slice and preview your model to check for any possible issues. As discussed in the previous practical applications, it's important to check the first layer of your print to ensure that all pieces are touching the build plate, as shown in *Figure 18.31*:

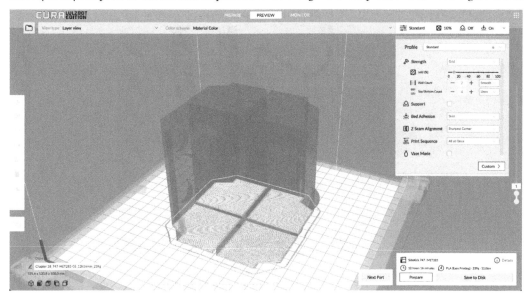

Figure 18.31: Inspecting the first layer of the print preview

After inspecting the model, we can save the *Gcode* file to manufacture these parts. You can then repeat these steps with the remaining pieces of the puzzle if you choose to manufacture pieces separately as I have.

Next, we can 3D print our parts, as shown in *Figure 18.32*:

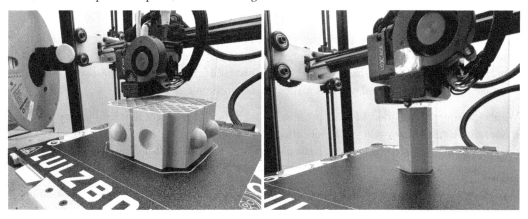

Figure 18.32: 3D printing the pieces of the puzzle

Once all the pieces have been manufactured, you can attempt to solve this puzzle! There shouldn't be any post-processing steps needed to prepare the pieces as supports were not needed for this design. Attempt to arrange the pieces in the correct order so that joints can interlock and so the pattern aligns to test fit the pieces, as shown in *Figure 18.33*:

Figure 18.33: Test fitting the puzzle pieces

Should the pieces fail to fit together, or if the pattern did not print correctly, return to your Tinkercad design and adjust as needed. Remember, designing projects like this from start to finish can sometimes be as much of a puzzle as the project itself!

Summary

While this practical application did not introduce any new tools that we have not used previously, it did challenge us to use our skills in 3D modeling in new ways for a different kind of project.

By copying, aligning, and grouping shapes, we were able to make puzzle pieces with similar features in unique ways. This was done without having to design each piece individually, but instead, by using templates that we created with shapes in Tinkercad. This was a far more efficient approach to designing a complex project like this, and one that makes the design simple and effective.

We also considered constraints for 3D printing through the design stages of the project by choosing shapes and features that would print more effectively by reducing the needs for supports. This made manufacturing the models easier using basic print settings and materials on lower resolution settings. As a result, the puzzle is not only a highly effective and efficient design, but also a successful product.

We will continue to expand on many of these concepts as we begin another practical application in the next chapter. However, for this one, we may need some more space as the next project will make things fly!

19
Designing and Assembling a Catapult

In this chapter, we will engage with another practical application to test our skills in 3D modeling and 3D printing, albeit with different challenges than what we have faced previously as we will create a mini-catapult! Like the last chapter, the pieces we create in this design will need to fit together, but as we strive to create a new product here, these pieces must combine a variety of fits and tolerances in order to function correctly. These activities will be broken down into the following topics:

- Designing the frame
- Creating an assembly system
- Creating the projectile system
- Manufacturing and prototype testing

Throughout this chapter, you will have the opportunity to continue to develop real-world skills in advanced 3D modeling techniques using Tinkercad as we revisit some concepts and tools introduced previously, such as the ruler tool and effective strategies for 3D printing production. By the end of this chapter, you will feel more comfortable in designing models so that they can be manufactured with 3D printing more efficiently, and you will also have a fun mini-catapult to play with too!

Technical requirements

As with the previous chapters, we will need access to a 3D printer to manufacture the parts created in this chapter. Additionally, you will need rubber bands to assemble your catapult after it has been manufactured. Editable example models of the catapult shown in this chapter can be found on Tinkercad at the following links:

- `https://www.tinkercad.com/things/20KuRl3NUfh-catapult-model-from-chapter-19`

- `https://www.tinkercad.com/things/anV9CHeENsr-simulated-catapult-model-from-chapter-19`

Designing the frame

Whenever you start a complex multi-part project like this one, it's always a good idea to plan out your designs before diving in. As discussed earlier in *Chapter 2*, this can be done through a variety of brainstorming and sketching techniques, with the most advanced option being to create a **technical drawing** like the one shown in *Figure 19.1*:

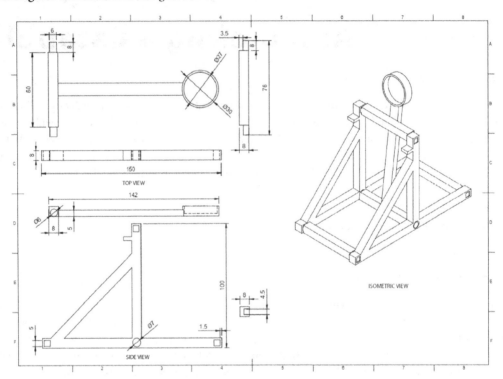

Figure 19.1: A technical drawing for a catapult

Designing the frame 343

The drawing shown in *Figure 19.1* is of the catapult we will make in this chapter. While this drawing was generated digitally, you could also sketch one by hand, which is sometimes quicker and more effective when you're planning your initial ideas. This drawing allows us to organize important dimensions, as well as get a sense of what this project entails as we can see the different parts and how they will fit together from **orthographic** and **isometric** perspectives.

As we start to model the frame for our catapult, we can continue to reference the drawing shown in *Figure 19.1* as necessary to check the dimensions throughout this process.

Starting with a Box shape

To start modeling, create a new Tinkercad 3D design and begin by bringing a **Box** shape onto the **Workplane**, as shown in *Figure 19.2*:

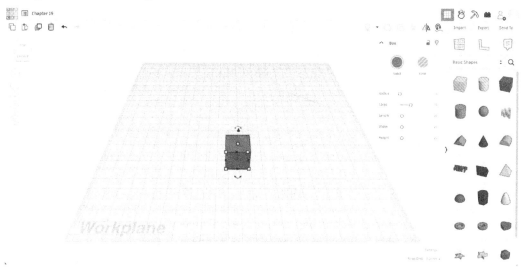

Figure 19.2: Starting a new 3D design in Tinkercad

After bringing the **Box** shape into your design, we can begin to model the frame pieces for the catapult based on the dimensions in the drawing shown in *Figure 19.1* for this project. You may find that entering these measurements is easier to do when using the **Ruler** tool, as introduced in *Chapter 6*. You can drag the **Ruler** tool onto the Workplane and align it with the front-left corner of the **Box** shape, as shown in *Figure 19.3*:

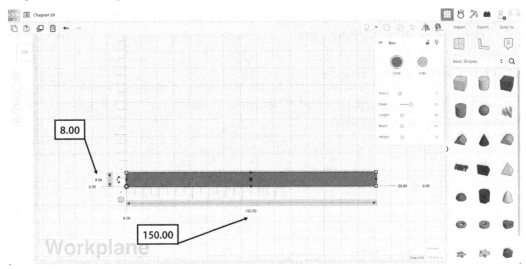

Figure 19.3: Dimensioning the box shape using the Ruler tool in an orthographic perspective

You may also find that creating the parts for the catapult is easier to do when looking at them from the top view in a 2D *orthographic* perspective, as discussed earlier in *Chapter 3* and as shown in *Figure 19.3*. After enabling the **Ruler** tool and changing our view, we can dimension the **Box** shape to be 150 mm x 8 mm in length and width, as shown in *Figure 19.3*. We'll dimension the height of this part later on.

Adding shapes to make our part

We can now finish modeling the frame by adding pieces to the **Box** shape through the following steps:

1. Copy the existing **Box** shape or add another one to create a center beam, perpendicular to the box already in the design, as shown in *Figure 19.4*:

Designing the frame | 345

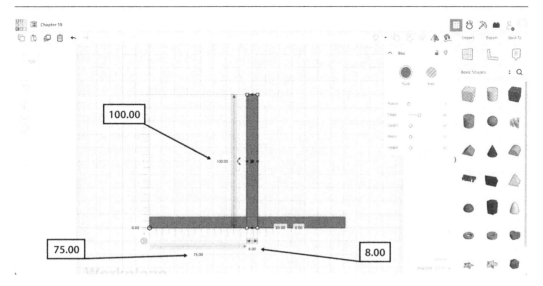

Figure 19.4: Creating a center post for the frame

This center beam should be dimensioned to be 100 mm x 8 mm in length and width, as well as centered by setting it to be 75 mm from the left side, as shown in *Figure 19.4*.

2. With the center beam selected, press the **Duplicate and repeat** button on the top toolbar to make a copy of this shape. Then, rotate the shape -45 degrees, as shown in *Figure 19.5*:

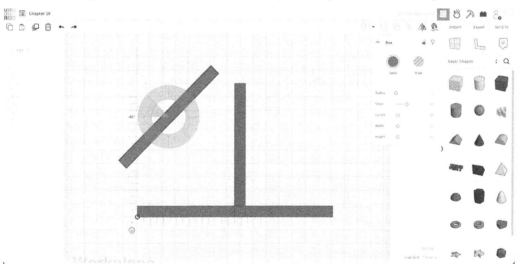

Figure 19.5: Rotating a copy of the center post shape

3. Next, move the angled **Box** shape created in the previous step so that it creates a truss-like connection between the bottom beam and the center beam, and is set inward by 6 mm from the left side, as shown in *Figure 19.6*:

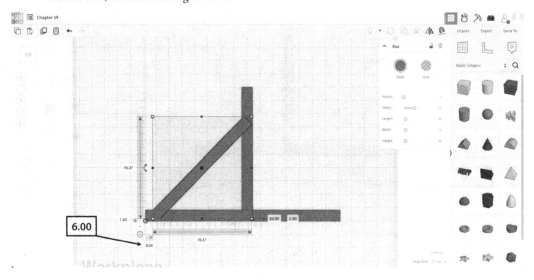

Figure 19.6: Creating a structural connection between the frame pieces

4. Lastly, select all three **Box** shapes used to create the frame and press the **Group** button to group these pieces together, as shown in *Figure 19.7*:

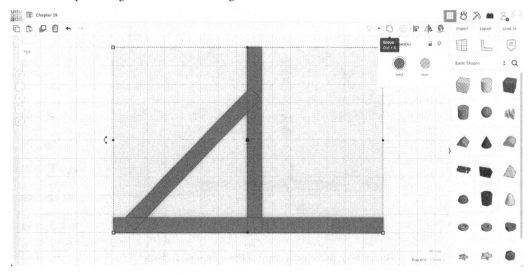

Figure 19.7: Grouping the frame pieces to make a solid shape

With the core structure of the frame now modeled, we can move on to create a system that can be used to assemble the catapult.

Creating an assembly system

There are, of course, countless ways to design and create a catapult – not only in regards to the type of catapult but also in how the parts of the catapult will fit together. To create a successful design of your own, you must consider the resources that will be used to assemble the catapult in your design. For example, we should consider the limitations of 3D printing, such as **tolerances** or **overhangs**, and design our pieces to cater to these limitations accordingly. We should also consider whether we want to use glue to assemble the parts, or design them in such a way that no adhesive would be needed.

In the following steps, we will create an assembly system that is not only structurally strong enough so that the catapult performs, but one that is also designed for easy manufacturing using lower-quality 3D printing techniques, as well as without needing glue later on.

Creating the assembly holes

The first step in making the assembly system will be creating holes placed along the frame piece designed in *Designing the frame* section:

1. To make printing the pegs that will fit in these holes easier, we will make them using square hole **Box** shapes, as shown in *Figure 19.8*:

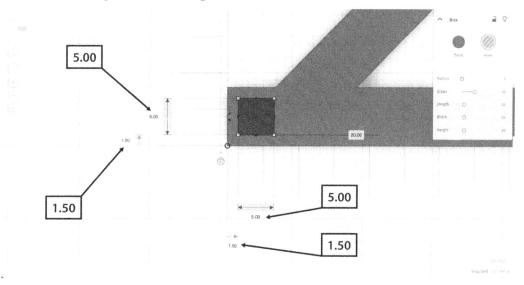

Figure 19.8: Placing the first assembly hole onto the frame of the catapult

With the **Ruler** tool still enabled, the first assembly hole can be dimensioned to be 5 mm in length and width and set to be 1.5 mm in from the bottom and left side of the frame, as shown in *Figure 19.8*.

2. We can then copy and move a second assembly hole **Box** shape so that it is positioned to be 1.5 mm in from the right side of the frame, as shown in *Figure 19.9*:

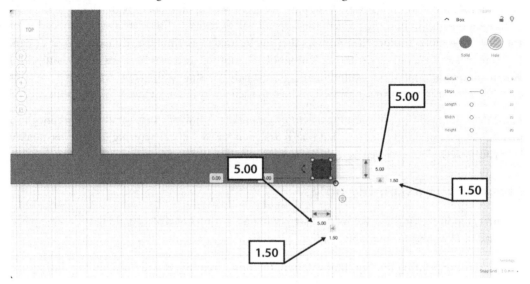

Figure 19.9: Making a second assembly hole on the frame of the catapult

It's important to use the same size box for all the assembly points so that they are universal. This will make for a more efficient and easier-to-use design later.

3. We can copy and move the assembly hole **Box** shape one more time and position a third hole at the top of the center beam so that it is 1.5 mm down, as shown in *Figure 19.10*:

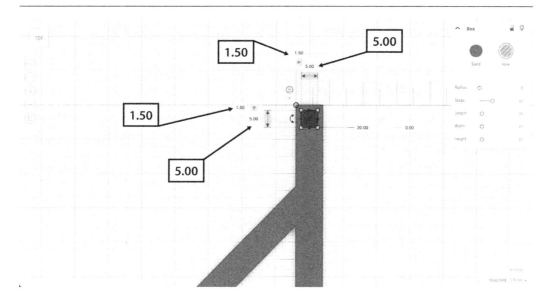

Figure 19.10: Adding a third assembly hole on the frame of the catapult

4. Lastly, we can select the frame shape and all three assembly hole shapes created and then **Group** them together to make a solid part, as shown in *Figure 19.11*:

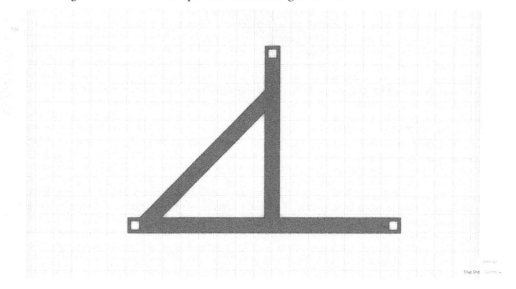

Figure 19.11: Grouping the hole shapes with the frame shape

With the holes created in the frame, we can now move on to creating structural pieces that connect the catapult.

Creating the Assembly Beams

Before continuing the assembly system, we will leave the orthographic view and switch to a 3D *perspective* view so that we can inspect the frame piece from a different viewing angle. We can also set the thickness, or height, of the frame piece as 8 mm, as shown in *Figure 19.12*:

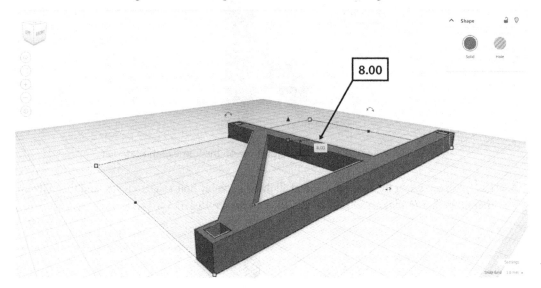

Figure 19.12: Adjusting the height of the frame from a 3D perspective view

Next, we need to make beams that will span across two frame pieces and connect the parts together, using the assembly holes made in the *Creating the assembly holes* section. These beams can be created through the following steps:

1. With the frame temporarily hidden or moved out of the way, we can add a new **Box** shape that is dimensioned to be 60 mm x 8 mm x 8 mm in length, width, and height, respectively, as shown in *Figure 19.13*:

Creating an assembly system | 351

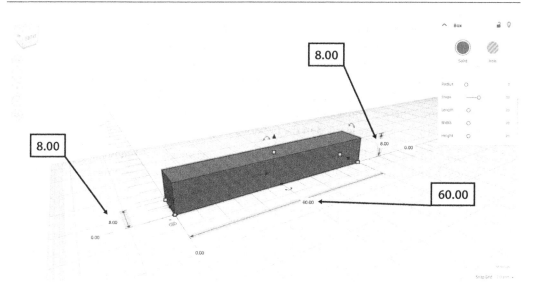

Figure 19.13: Adding a new box shape for the structural beam pieces

The length of 60 mm will set the overall width of our catapult, as this will be how far apart the two frame pieces are from one another when assembled. The dimensions of 8 mm are set to match the ones used when creating the frame, as depicted in the drawing shown earlier in *Figure 19.1*.

2. We can then copy or create a new **Box** shape and dimension it to be 76 mm x 4.5 mm x 4.5 mm in length, width, and height, respectively, as shown in *Figure 19.14*:

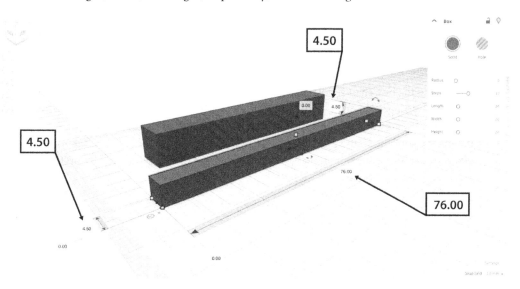

Figure 19.14: Creating a new box shape for the structural beams

This box shape will be used to create pegs that fit into the assembly holes we made earlier. By setting the width and height of this new box shape to be 4.5 mm, we created a tolerance gap of 0.5 mm between this box and the hole boxes we made earlier in the frame. As discussed in *Chapter 12*, this is within the tolerance to create a **Transition Fit** for typical **FFF**-style 3D printers using a **PLA** filament, which should make assembling these pieces easy to do without needing any adhesive. You should adjust the size of this box based on the tolerances of your printing resources, as we learned in the previous chapter.

3. We can then use the **Align** tool to position the two box shapes within one another. The box shapes should be centered in the X and Y axes but flush with one another in the Z axis, as shown in *Figure 19.15*:

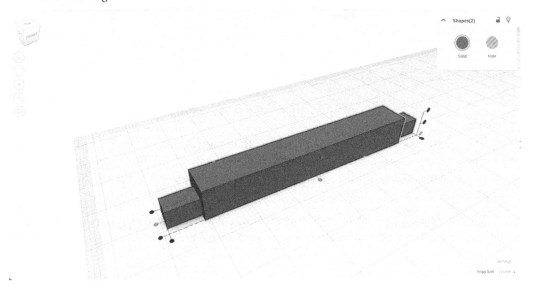

Figure 19.15: Aligning the box shapes for the structural beams

By keeping the box shapes flush on one side, we avoid making an **overhang**, allowing for an easier and stronger 3D print. Also, note that because one box shape is 16 mm longer than the other, there is now an 8 mm peg protruding on either end of the beam piece, which matches the thickness of the frame piece we set in the *Creating the Assembly Beams* section. Once the two box shapes have been aligned, as shown in *Figure 19.15*, you can **Group** these shapes together to finish this part.

After creating the structural beam part, we can make two copies of it so that there are three beams in total. We can then bring the frame part back into the design and make a copy of it so that we have enough pieces to create a mock assembly, as shown in *Figure 19.16*:

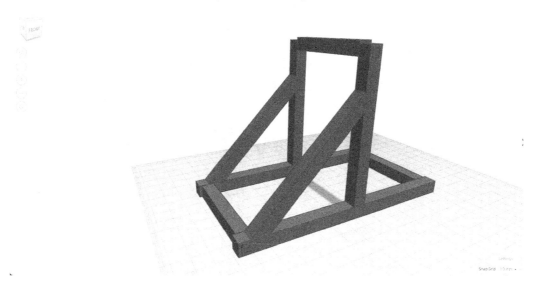

Figure 19.16: Assembling the catapult using the assembly components

This assembly not only allows us to test to see how the created parts will fit together; it also allows us to better visualize what this prototype would look like after manufacturing and adjust the design of our parts as needed. With the basic structural components now complete, we can move on to creating the projectile launching system that will allow this catapult to really perform!

Creating the projectile system

Of course, catapults are designed to do one thing – hurl a projectile some distance. The traditional catapult we are making in this project will do this by rotating a kind of bucket arm around an axle, propelled by the elastic tension of a rubber band.

Completing the catapult frame

After disassembling the pieces of our catapult from the mock assembly done previously, we can change our view so that we are once again looking at the frame of our catapult from a 2D *orthographic* view. This will allow us to easily start the projectile system by finishing the frame part through the following steps:

1. Unlike the structural beams that will hold the frames of the catapult together, the base of the projectile arm will need to be rounded for it to pivot smoothly. To do this, we can create a **Hole** in the frame of our catapult using a **Cylinder** shape that is 7 mm x 7 mm in length and width and centered, as shown in *Figure 19.17*:

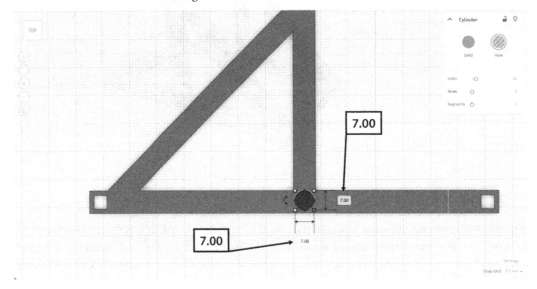

Figure 19.17: Creating a hole for the projectile arm to pivot in

2. The arm will move due to the energy harnessed by stretching a rubber band around it. To do this, we need to create attachment points for the rubber band, which can be done by adding a **Box** shape to the top of the catapult, as shown in *Figure 19.18*:

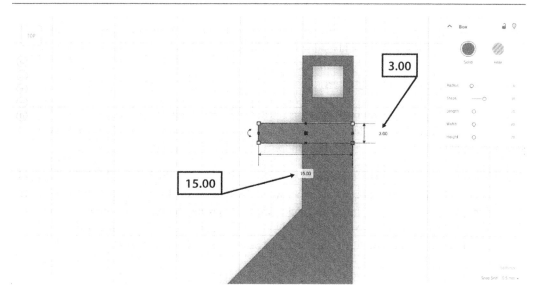

Figure 19.18: Adding attachment points for the rubber band

The dimensions for this **Box** are less critical than the other ones we've considered thus far, but it's important that this part is big enough for a rubber band to loop around. I've dimensioned my **Box** to be 15 mm in length, which protrudes 7 mm outward toward the front of the catapult, and I've made this shape 3 mm in height to offer some needed rigidity.

3. We can then change our view back to a 3D perspective view and adjust the added shapes to be the right height to match the frame of the catapult. We can then group these shapes together to finish the frame, as shown in *Figure 19.19*:

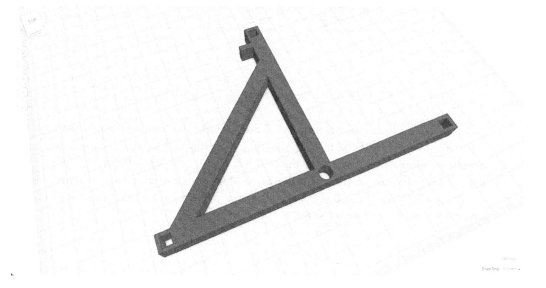

Figure 19.19: Completing the frame of the catapult

Making the projectile arm

With the frame part now completed, we can move on to making the projectile arm, which will pivot and move using the features we've just added to our design.

Making the projectile arm

The arm of the catapult is an important part, as it will need to fit within the parts we've already created while also supporting a projectile that has yet to be made. After moving or hiding the frame part so that it is out of the way, we can create the arm for our catapult through the following steps:

1. First, we need to create a **Cylinder** shape that will fit within the hole in the frame we made previously, spanning between the frames when assembled. As shown in *Figure 19.20*, I created a **Cylinder** that is 6 mm in width and height, which will create a looser **Clearance Fit** (as discussed in *Chapter 12*), allowing for the arm to pivot freely. I've also made this **Cylinder** shape to be 76 mm in length, which matches the structural beam pieces we made earlier.

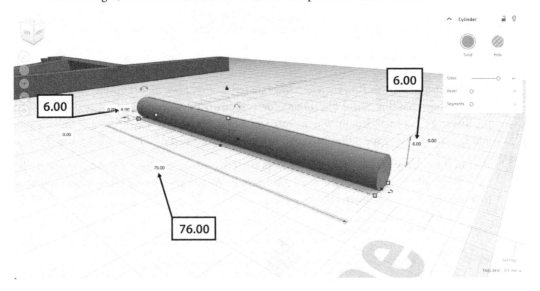

Figure 19.20: Starting the arm with a cylinder shape

2. As discussed in *Chapter 11*, printing a rounded part like the cylinder we created in the last step would be rather difficult to do successfully. To mitigate this, we are going to place this cylinder in a **Box** shape so that only the ends will be rounded, while the rest of the part will have a flat base for easy printing. This **Box** can be dimensioned to match the structural beams we made earlier so that it is 60 mm x 8mm x 8 mm in length, width, and height, respectively, and centered with the **Cylinder** shape, as shown in *Figure 19.21*:

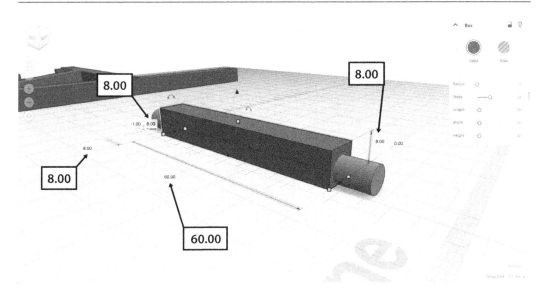

Figure 19.21: Adding a box shape around the cylinder shape for the arm

3. Next, we can add another **Box** shape that will create the main part of the projectile arm, and this arm needs to be long enough to reach over the topmost structural beam when assembled. As this was set to be 100 mm previously, we can make this new **Box** shape for the arm 120 mm in length, 10 mm in width, and 8 mm in height, as shown in *Figure 19.22*:

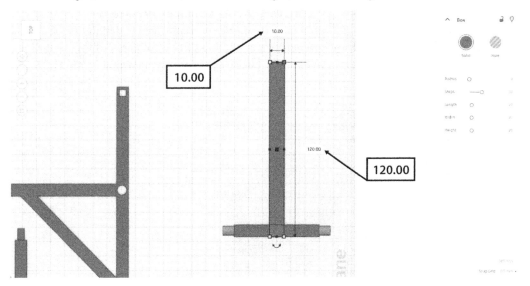

Figure 19.22: Adding a box shape to create the arm

4. We now need to create some sort of bucket or basket for the projectile to fit in. If you haven't already done so, consider what you want your catapult to launch. Will it be a ping pong ball? Or marshmallows? Or something that has been 3D printed? We'll have the option to make 3D projectiles later, but for now, you can add a bucket-like part using **Cylinder** shapes that will support the size of your chosen projectile, as shown in *Figure 19.23*:

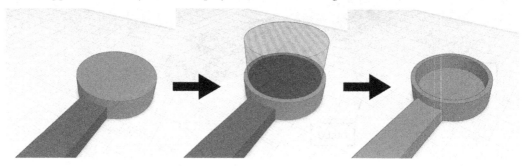

Figure 19.23: Creating a bucket on the projectile arm

Once all parts of the projectile arm have been added, you can **Group** all the shapes together to finish this part before moving on to the next step.

Making projectiles

As discussed in the previous section, you could choose from a wide range of objects for your catapult to hurl across your living room. But if you want, you can also use Tinkercad to design unique objects for use as projectiles. There are a lot of different shapes to choose from, but personally, I like the look of the **Icosahedron** shape when it's moved into the bucket of my arm, as shown in *Figure 19.24*:

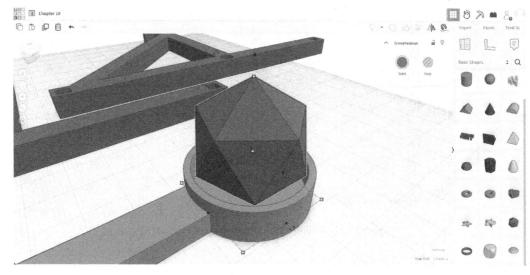

Figure 19.24: Testing out shapes to use as a projectile

We can use the bucket part of the arm that was made previously to scale our chosen projectile shape, as shown in *Figure 19.24*. For the projectile part to print successfully, you should ensure it has a flat base, as discussed previously in *Chapter 11*. This can be done by using a **Hole** shape to cut the bottom off your projectile shape, as shown in *Figure 19.25*:

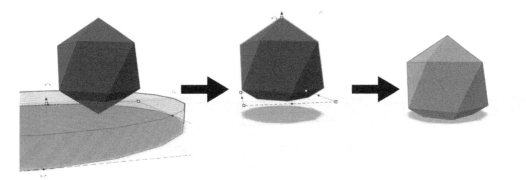

Figure 19.25: Creating a flat base on the projectile shape

With all the parts for the catapult projectile system now created, we can move on to manufacturing and testing this project!

Manufacturing and prototype testing

Before we move into exporting our catapult design for manufacturing, it would be good practice to do one more test fit of all your parts to ensure that they will assemble properly after manufacturing. Also, while it's not required, this would be a good opportunity to use Tinkercad's **Sim Lab** feature to simulate the functionality of your catapult, as shown in *Figure 19.26*:

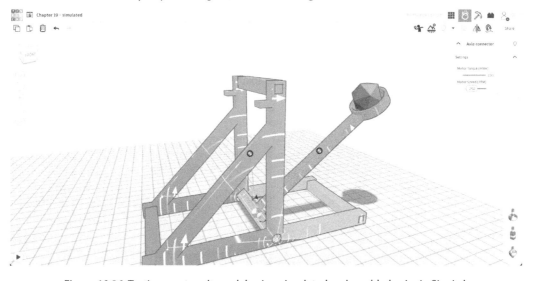

Figure 19.26: Testing a catapult model using simulated real-world physics in Sim Lab

As initially introduced in *Chapter 1*, Sim Lab allows you to apply real-world physics to your 3D designs, which may be a useful prototyping and testing tool for a project such as this. If your arm wasn't long enough, or if the frame was a bit unstable, performing a test simulation in **Sim Lab** may allow you to identify and correct some of these issues before moving into production, just like how industry professionals use CAD to test the products we buy and use every day!

Once all parts appear to be ready for manufacturing, we can move on to arranging them so that they are optimized for production using your 3D printer, as discussed in *Chapter 11*. You can first adjust the size of your **Workplane** to match the bed or build platform of your 3D printer, and then arrange all the parts for your catapult to fit, as shown in *Figure 19.27*:

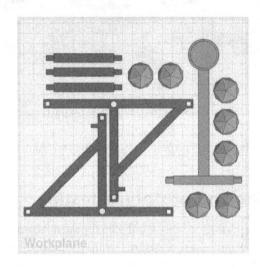

Figure 19.27: Arranging the parts of the catapult for production

It's important to ensure that all parts are flat on the **Workplane** and that they are not floating or angled slightly, as discussed in *Chapter 13*. Once this has been checked, you can press the **Export** button in the top-right corner of the design window to download or share your 3D models for production, as we did in the previous chapters.

If you are printing these parts yourself, the next step is to import the design files into the **CAM** program for your 3D printer, as shown in *Figure 19.28*:

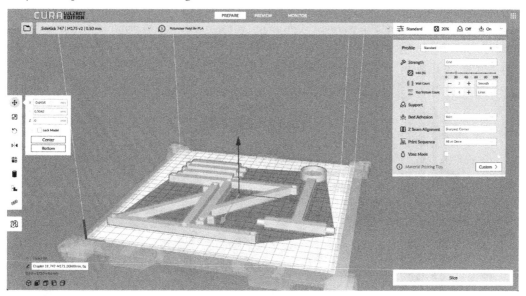

Figure 19.28: Preparing 3D models for production using Cura LulzBot Edition

As shown in previous chapters, I've imported the model for the frame of my catapult into *Cura LulzBot Edition*, which is the **slicer** program for my 3D printers. I've chosen to export the frame and projectile parts separately so that I can print them using different materials on different printers, which is not essential for this project. Before moving on to printing these parts for our catapult, there are a few important settings to consider:

- **Material**: Like the last few projects, we could use PLA as an optimal material choice for novelty projects such as this catapult. If you wanted, you could choose to use a soft material such as TPU for the projectiles, which is something we will look more at in the next chapter.
- **Quality**: A standard print quality should work for the catapult frame pieces, as we've accounted for our printer's tolerance in the design. Depending on the size and shape of the projectile pieces, you may want to use a higher quality to print those parts if they are smaller or of finer detail.
- **Infill density**: A minimum infill density of 20% should be used to print the frame pieces as there will be stress placed on these parts during use.
- **Supports**: Because we designed the beams to be square and without overhangs, no support material should be needed if all parts have been placed flat onto the Workplane before exporting.

After considering these settings and slicing your model, don't forget to preview the first layer to ensure that all parts are arranged correctly, as we did in previous chapters. You can also repeat these steps for the remaining parts of the catapult if you choose to export them separately as I have. Once everything looks as it should in your CAM program, you can move on to printing the parts of your catapult, as shown in *Figure 19.29*:

Figure 19.29: 3D-printing the parts of the catapult frame on an FFF-style 3D printer

Once the pieces have been 3D-printed together, or separately as I have done, we can move on to assembling and testing our catapult designs through the steps shown in *Figure 19.30*:

Manufacturing and prototype testing 363

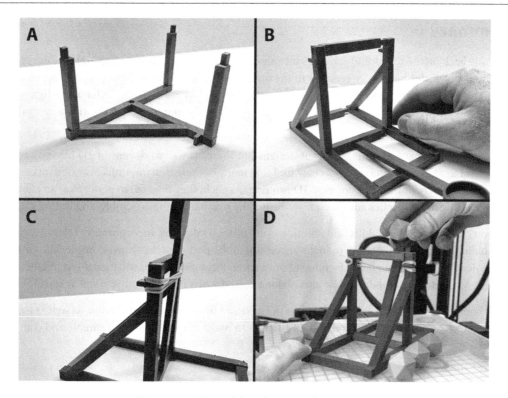

Figure 19.30: Assembling the catapult prototype

As shown in *Figure 19.30*, we can assemble and test the catapult by following the following steps:

A. With one of the frame pieces lying on a flat surface, install all the structural beam parts into the assembly holes.

B. Before attaching the second frame piece, insert the projectile arm into place between the two frame pieces. Once all pieces are aligned, press the second frame piece onto the pegs of the structural beams.

C. Wrap one or two rubber bands around the arm when it's in an upright position using the attachment points printed at the top of the frames.

D. Load your projectiles, and carefully test your catapult prototype!

Practice safety first with this project! Make sure you handle your catapult with care and launch your projectiles responsibly. Also, as you test the functionality of this prototype, consider how the design could be improved with a longer arm, additional rubber bands, or perhaps even with a different approach such as making a trebuchet or ballista-style device instead. Remember, with the skills we've gained throughout this chapter and earlier in this book, anything is possible as you continue to develop your designs and engage with rapid prototyping techniques!

Summary

Looking back on the topics and activities introduced in this chapter, there are several takeaways that we should consider before progressing to the final chapter of this book. This practical application allowed us to strengthen our skills using tools we've seen before, such as the **Ruler** and **Align** tools, and it also allowed us to explore new challenges when designing prototypes with successful 3D printing strategies in mind.

On the web, there are countless catapult designs out there, as well as dozens of 3D files you could download to 3D-print. However, you may find that many of these designs, unlike the one we created in this chapter, are not suitable for easy 3D printing. As such, they may fail to perform as an effective product, causing headaches and challenges during the manufacturing, assembly, and testing stages.

As you continue to apply skills and abilities in creating 3D designs, incorporating tolerances, and combining different types of fits to make successful multi-part models, always consider the entirety of the design process from brainstorming to production throughout. This will increase your efficiency in designing advanced models while also making your designs more effective from the start.

We'll continue to consider these concepts in the next and final chapter of this book, as well as look at some new 3D printing resources that may allow us to make one of our most desirable and complex projects – a 3D printed phone case!

20
Prototyping a 3D-Printed Phone Case

We now find ourselves in the final chapter of this book, as well as faced with another practical application to test our skills and knowledge gained over the past 19 chapters. In this chapter, we will be designing and manufacturing one of the most complex and desirable 3D-printed projects, a custom 3D-printed phone case.

At first, this project might not seem too challenging, or even as complex as some of the ones we have done previously. But you may find that creating a phone case pushes the limitations of both Tinkercad and the typical 3D printers that you might find at home. These concepts and challenges will be broken down through the following topics:

- Acquiring the dimensions
- Modeling the phone
- Making the case
- Adding aesthetic and ergonomic features
- Manufacturing with specialty materials

As you progress through this chapter, you will have the opportunity to apply many of the skills and tools that we have been learning about through a new real-world challenge that offers an exciting opportunity to make something you could use every day. This will also allow you to gain additional understanding and abilities in the methods and steps needed to prototype real products as we strive to improve the functionality and performance of our designs later in this chapter. By the end of this chapter, you will feel confident in planning out a project, as well as being able to utilize Tinkercad to develop effective 3D models for 3D printing in even the most challenging circumstances!

Technical requirements

To manufacture the project created in this chapter, you will need access to a 3D printer that can print flexible materials, such as TPU. You may also need measuring tools, such as digital calipers, as initially introduced in *Chapter 2*. An editable model of the phone case shown throughout this chapter can be accessed on Tinkercad at https://www.tinkercad.com/things/hQLnjnzAGMh-phone-case-model-from-chapter-20.

Acquiring the dimensions

As discussed in *Chapter 2*, the first step in starting a complex project like this is to acquire all the necessary dimensions needed to design our models. As the case we are designing in this chapter needs to fit around an existing product to work successfully, we need to take crucial dimensions from that device so we know where to start.

As cases and mobile device accessories are such readily available products, it is very likely that you can find the dimensions for the device you are looking to design a case for on the web. Try searching for phone model dimensions, replacing phone model with your actual device model, and see whether you can find a technical drawing like the one shown in *Chapter 2*.

You also can take these measurements yourself using a ruler or calipers, as shown in *Figure 20.1*:

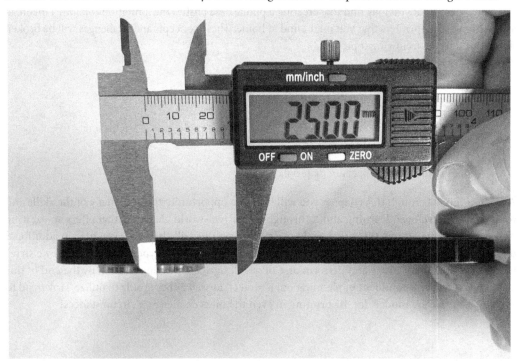

Figure 20.1: Using digital calipers to measure the buttons on a phone

Gaining skills in using measuring tools to record measurements as shown in *Figure 20.1* is important, and universal for any type of project you might find yourself creating. Even if you can find existing technical drawings for your mobile device on the web, some key dimensions might be missing that you need to be able to determine effectively.

While this chapter shows the steps needed to create a phone case for my own device, you could implement the same steps and strategies to make a case for a different type of device instead. For example, perhaps you could take measurements to create a case for your wireless headphones, mobile gaming device, key fob, or tablet instead. The concepts and skills that are to be covered in the following topics would be applicable to any mobile device such as these. Once you have found or acquired your key measurements, we can move on to modeling the device we are designing a case for in Tinkercad.

Modeling the phone

As we have learned, there is never one way to approach any challenge. As such, there are many ways to attempt to design a phone case in Tinkercad. I find the most effective way to create the case is to actually model the phone we've just measured first. If we were working in a parametric CAD program, as discussed in *Chapter 4*, I would probably consider starting right with the case, but as Tinkercad's tools and **CSG** features cater to modeling around shapes, that shape can be the device we wish to protect with our case design.

We can start by creating a new Tinkercad 3D design and bring a **Hole Box** shape onto the **Workplane**, as shown in *Figure 20.2*:

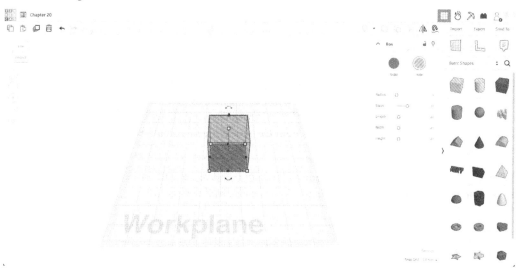

Figure 20.2: Starting with a Hole Box shape

The model of our phone will later be used to cut the cavity in the case we are creating, which is why we are starting with a **Hole** shape rather than a **Solid** shape, though of course, it's easy to toggle between these options, as we learned in previous chapters.

Then, as learned previously in *Chapter 6*, we can use the **Ruler** tool to dimension the **Box** shape to match the overall size of the device we measured in the previous section, as shown in *Figure 20.3*:

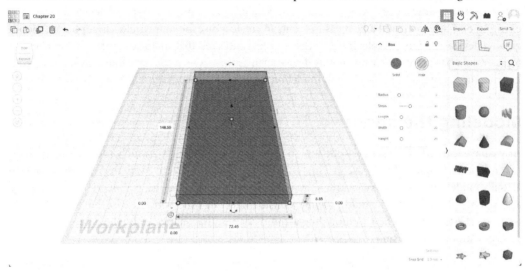

Figure 20.3: Adding dimensions to the design using the Ruler tool

I won't be focusing too much on specific measurements in this chapter as it is unlikely that you and I are designing for the same device. But you can see that I've dimensioned my box shape's width, length, and height to match my phone in *Figure 20.3*. I've also added 0.5 mm as a **tolerance** to the dimensions that I recorded previously to create a **transition fit** between my phone and case design, as we learned in *Chapter 12*. The tolerances you add to your measurements will of course vary based on the printer and printing material you plan to use later in the manufacturing stages of this project. But you may find that you can utilize tighter tolerances than we have done in previous chapters if you plan to use flexible printing materials as I will be.

With the overall measurements incorporated into our design, we can move on to adding more key features to this model so that it better reflects the device we are designing for.

Rounding the corners

It's very likely that your device has rounded edges or rounded corners rather than perfectly sharp and square ones. As such, we need to design our case to account for these features so that we can ensure a proper fit.

Rounding all the edges of the **Box** shape is rather easy to do, as you can simply adjust the **Radius** shape parameter, as learned previously in *Chapter 11*. But placing a radius or corner **fillet** on the individual corners is a bit more challenging. We can do this by adding a couple of shapes and making a corner-cutting template, as shown through the following steps:

1. First, add a new **Box** shape set to **Solid** and a **Cylinder** shape set to **Hole**, and align the shapes so that the **Box** intersects one quadrant of the **Cylinder**, as shown in *Figure 20.4*:

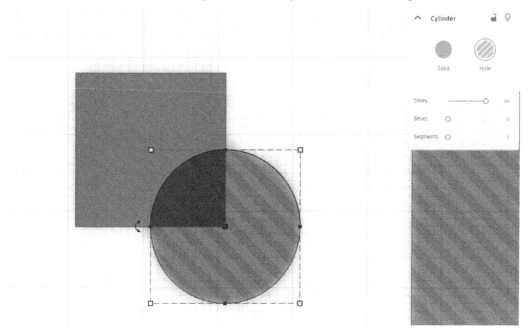

Figure 20.4: Aligning a Box and Cylinder shape to create a cutting template

2. Next, group these two shapes together to create the template shape, as shown in *Figure 20.5*:

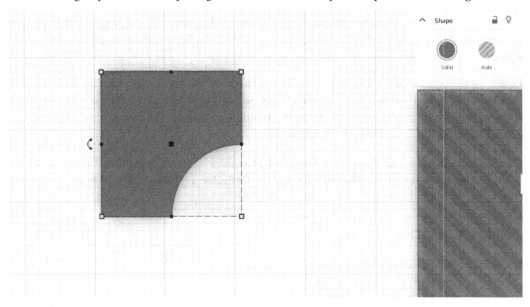

Figure 20.5: Creating the corner template shape

3. Scale the template shape to match the corner radius of your device based on the measurements you took earlier, and move and align copies of the corner template with the **Box** shape for your device, as shown in *Figure 20.6*:

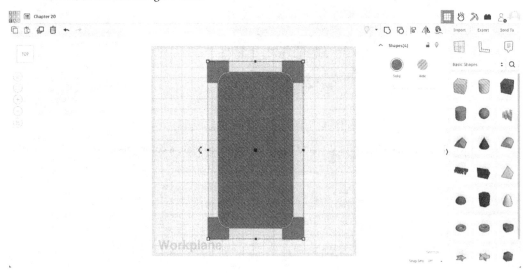

Figure 20.6: Aligning the corner template shapes with the Box shape for our device

4. Set the corner template shapes to **Hole** and the **Box** shape for the phone to **Solid**, then group the shapes together, as shown in *Figure 20.7*:

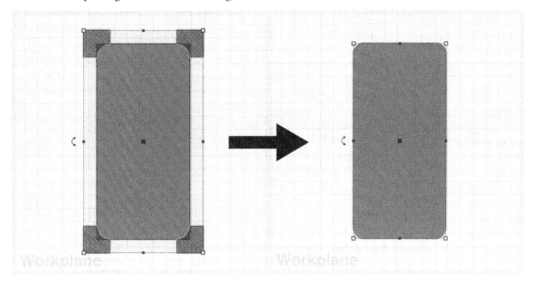

Figure 20.7: Cutting rounded corners into the Box shape for our mobile device

With the rounded corners cut to match our device, we can set the **Box** shape for the phone back to being a **Hole** shape rather than **Solid**. We can then move on to modeling the different ports and buttons on our device in the next steps.

Adding the ports and buttons

Whichever device you have chosen to design a case for, it's likely to have several connection ports and buttons along the sides or back face that we need to design around. To do this, we are going to add **Hole** shapes that will later cut into the model of our case through the following steps:

1. Starting with one side, I've changed my view to a 2D *orthographic* view to make working on the flat surface a bit easier to do, as discussed in previous chapters. I've then added a **Hole Box** shape dimensioned to match the button on the right side of my phone, and positioned it the correct distance from the top edge using the **Ruler** tool, as shown in *Figure 20.8*:

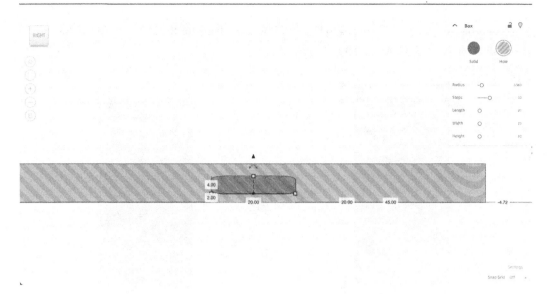

Figure 20.8: Adding a Box shape for the first button on the device

As discussed in *Chapter 6*, you can change the direction and the axis of the **Ruler** tool by clicking on its origin point. This makes aligning and moving shapes around the edges of our model easy and accurate, as shown in *Figure 20.8*. As you add the **Hole Box** shapes for the buttons and ports, the distance they protrude from the main **Box** shape is not critical. It's better to have them protrude more, as this will ensure they cut through the case shape we'll be adding later. You also may want to consider adjusting the **Radius** of these shapes to round the edges, which may not only better match your device but also reduce overhangs as well.

2. We can then repeat *Step 1* for any other sides that have buttons, as shown in *Figure 20.9*:

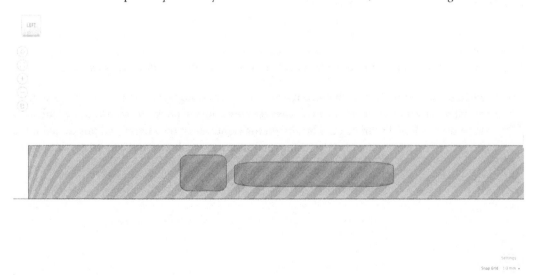

Figure 20.9: Adding Box shapes for buttons on the left side of the device

3. We can then repeat *Step 1* for any ports that might be placed along the back or bottom edge of the device, as shown in *Figure 20.10*:

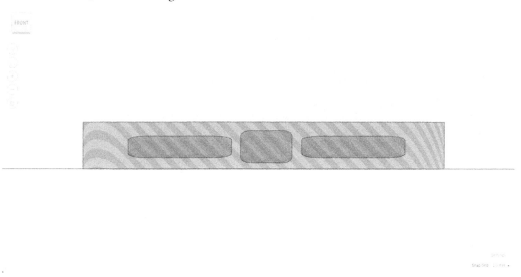

Figure 20.10: Adding Box shapes for ports on the bottom side of the device

Continue to reference the dimensions you took earlier in the *Acquiring the dimensions* section and don't be afraid to take additional dimensions or double-check them as you work!

4. Lastly, we can add **Box** or **Cylinder** shapes to the bottom of the device if there are any additional ports or buttons to account for, or if there are camera lens protrusions as I have on my device. I have also utilized the same corner-cutting method implemented in the *Rounding the corners* section to create the **Hole Box** shape needed for my phone's camera, as shown in *Figure 20.11*:

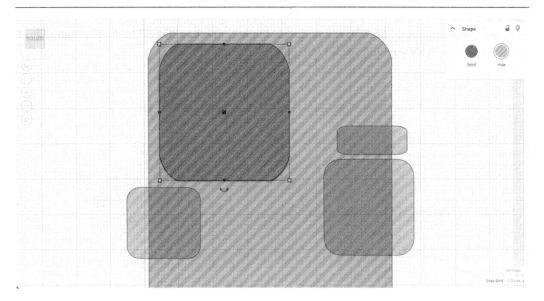

Figure 20.11: Adding shapes on the back of the device

With all the protruding shapes for ports and buttons in place, we now have our phone model, as shown in *Figure 20.12*:

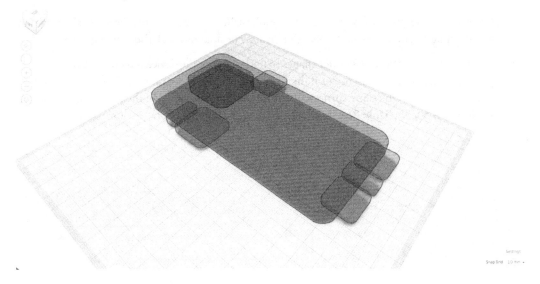

Figure 20.12: The starter model of a mobile phone

This model, which has been accurately dimensioned based on the device we are designing for, will now serve as an accurate starting point to design a case around in the next section.

Making the case

As we have now created a dimensionally accurate model of our mobile device, we have a collection of shapes to work from as we move on to making the case.

Making a copy of the phone model

The first thing we want to do is to make a copy of the phone model, as shown in *Figure 20.13*:

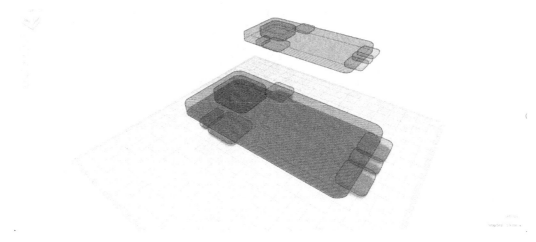

Figure 20.13: Copying the phone model to start the case

We can place one copy off to the side of the Workplane area, as shown in *Figure 20.13*, or hide it, as discussed in *Chapter 4*. With the other copy, we can set the main **Box** shape of the phone model to **Solid**, and then dimension this shape to be slightly larger than the **Hole Box** shape we had made to represent our phone, as shown in *Figure 20.14*:

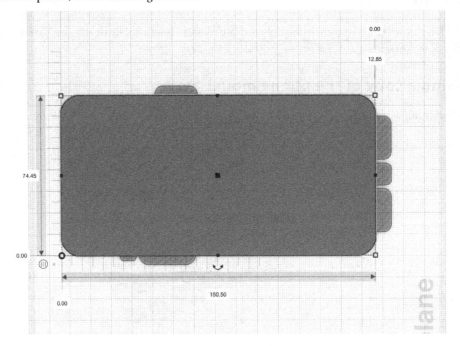

Figure 20.14: Modeling the case body

This **Box** shape will serve as the body for our case, and the difference in size between the larger **Solid** box and the original **Hole** box will equal the wall thickness of the case. For example, if I want my case to have walls that are 2 mm in thickness, I will need to add 4 mm to the length, width, and height of this box, as shown in *Figure 20.14*.

Creating the ports and buttons

We then need to cut the ports and buttons into the case body, but there are a couple of ways to do this. You could choose to group the **Hole** shapes with the case body box to cut holes for both the ports and the buttons, as shown in *Figure 20.15*:

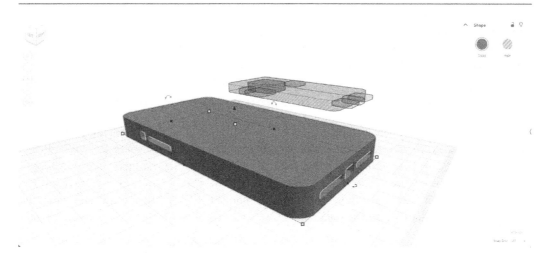

Figure 20.15: Option 1 – cutting holes for ports and buttons in the case body

Alternatively, you might want to create contact points for the buttons instead of cut-outs. To do this, turn the boxes for the buttons into **Solid** shapes and adjust how much they protrude from the case body so that they create rounded buttons with minimal overhangs, as shown in *Figure 20.16*:

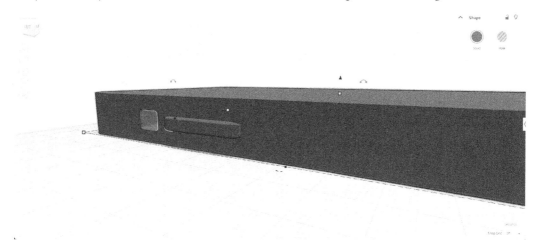

Figure 20.16: Option 2 – creating contact points for the buttons and holes for the ports

Either the full cut-out option shown in *Figure 20.15* or the contact point button option shown in *Figure 20.16* should work well, as long as you keep your overhangs to a minimum and round your openings, as we learned about in *Chapter 11*.

Using the copy to create the opening for the phone

With the port and button shapes grouped to the case body, we can bring the copy of the phone model back into our design. We can remove the boxes used for the ports and the camera as these have already been cut into the case, but we might need to leave the button boxes, depending on how we chose to model the buttons on our case. For example, if you chose to cut openings for the buttons as shown in *Figure 20.15*, you can delete the button boxes from the phone model and then center and group the model with the case body, as shown in *Figure 20.17*:

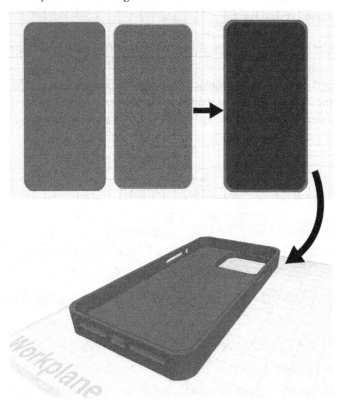

Figure 20.17: Aligning the phone model shape into the case body for cut-out buttons

Alternatively, you may want to keep the button boxes as holes on the phone model if there are physical contact points for the buttons, as shown in *Figure 20.16*. We would need to reduce how much the button boxes protrude so that small notches can be cut on the inside of the case without creating holes, as shown in *Figure 20.18*:

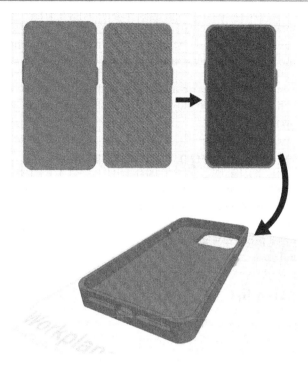

Figure 20.18: Aligning the phone model shape into the case body with notched buttons

For either design option, you may find that using the **Align** tool, as was introduced in *Chapter 7*, is a little difficult to do with so many extruding shapes. Instead, align the top edges of the phone model with the case model first, then use the **Snap Grid** feature introduced in *Chapter 2* to accurately move the phone model into the center of the case, as shown in *Figure 20.19*:

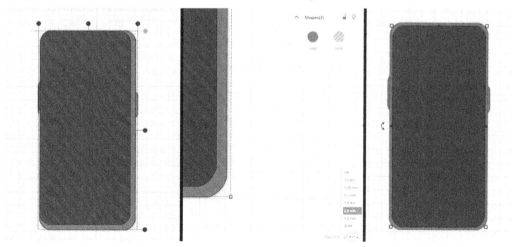

Figure 20.19: Steps to align the phone model with the case model

For example, I know I dimensioned my case to have a wall thickness of 2 mm. Once I've set **Snap Grid** to **2 mm**, I could easily use the arrow keys on my keyboard to position the phone model perfectly in the middle of the case model, despite the extruded pieces, as shown in *Figure 20.19*.

Once we've decided how we would like to create the buttons on our case and have grouped the necessary shapes together, we are ready to move on to the final design stages as we consider how this basic case can be improved.

Adding aesthetic and ergonomic features

Phone cases come in countless varieties, from ones that are flashy to others that are rugged. There are endless criteria that can be considered and incorporated into your model as you work to complete this real-world project. Two things that should not go without consideration are things that will make the case perform better as it protects your device, and design aspects to personalize this custom case to fit your own needs and style.

Adding a protective lip

The first thing we are going to add is a lip that will protect the screen of our phone from the top of the case. As we do this, we need to keep the limitations of 3D printers in mind and work to avoid creating any excessive overhangs, especially if we are going to manufacture this model using flexible materials. Flexible materials such as TPU are more difficult to print, reducing the overall quality that we can expect in our model. Additionally, removing support material, as discussed in *Chapter 11*, is much more difficult to do with flexible materials, so avoiding the need for support material in our case design is the best approach whenever possible.

As such, we can go through the following steps to add a lip to our case without creating an overhang that needs support material to print properly:

1. Start by making a copy of your case model and hide or move one copy off to the side, as shown in *Figure 20.20*:

Figure 20.20: Copying the case model to begin to make the lip part

2. **Ungroup** the case model and delete all shapes for the ports and buttons so that only the outermost box and inner box for the phone remain, as shown in *Figure 20.21*:

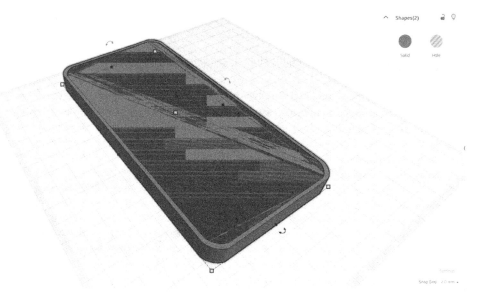

Figure 20.21: Removing the shapes for the ports and buttons to create the lip part

3. Adjust the dimensions of the **Hole** shape for the phone so that it is small enough to cover the edges of your device, but not cover the screen. This should only be about 1 or 2 mm smaller to avoid making a large overhang, as shown in *Figure 20.22*:

Figure 20.22: Adjusting the Hole shape so the device screen can still be visible

4. Adjust the **Hole** shape so it is taller than the **Solid** shape, **Group** the two shapes, and then flatten the grouped shapes to create the lip. The thickness of the lip should be about 2 mm, or match the wall thickness which you set earlier, as shown in *Figure 20.23*:

Figure 20.23: Creating the lip

5. Lastly, bring the case copy back into the design space, then align and group the lip part onto the surface of the case part, as shown in *Figure 20.24*:

Figure 20.24: Adding the lip to the case part

This lip should now work to not only hold your device in the case better but also protect it should it fall onto its screen or an edge. We can then move on to personalizing our case further through some more design features.

Adding unique design features

How would you like to personalize your case design? Do you want to add text features like we did in *Chapter 5*? Or perhaps custom artwork like we did in *Chapter 8* instead? As this is a case for a mobile device, you might also want to consider adding a lanyard of some kind, or an ergonomic grip to hold onto, such as the ones we created in *Chapter 17*. There are countless ways to customize your case so that it is personalized to your own taste and needs, while also creating a design that can successfully be manufactured using 3D printers.

As you enhance your case design, remember to keep overhangs in mind and keep flat surfaces touching the build plate, as discussed in *Chapter 13*. This is even more important when using specialty print materials such as flexible TPU, as this tends to require a better first layer to build from than most rigid materials might.

For my case, I have chosen to add a simple pattern that cuts into the back of my case, as shown in *Figure 20.25*:

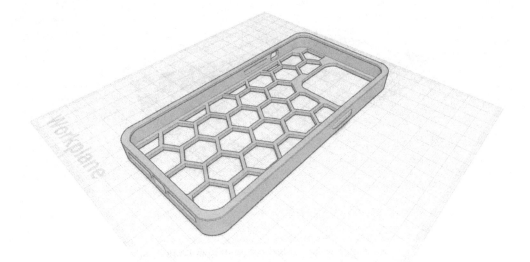

Figure 20.25: Adding a patterned design to the back of the phone case

The pattern design shown in *Figure 20.25* allows me to personalize my case without adding any overhangs. While the surface area touching the build plate has been reduced, only flat surfaces are still contacting the **Workplane**, which should allow for a successful print. I've also adjusted the color of my case to preview what it might look like using the material I will be printing with.

Once you've adjusted your case design as desired, we can move on to the final stages of this project as we prepare it for manufacturing with specialty materials!

Manufacturing with specialty materials

As we've done in previous chapters, we can now move on to exporting our model so that we can manufacture it using 3D printing technology. For the case we've designed in this chapter, however, we are going to do things a bit differently as we will be using specialty materials to manufacture a flexible silicon-like phone case.

But we've already completed the first step to manufacture a design using specialty materials, which was considering the limitations of the materials we are using in the design stages. As we've worked to include tighter tolerances in our dimensions, rounded openings and extrusions, minimum overhangs, and flat surfaces contacting the build plate, our design should be suitable for production using flexible materials.

Once we've exported our design from Tinkercad, as covered in *Chapter 14*, we can import the model into the **CAM** program for our 3D printer, as I have done for mine in *Figure 20.26*:

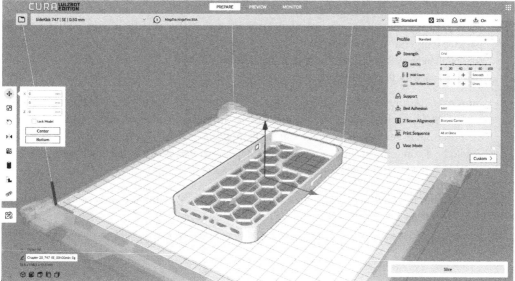

Figure 20.26: Importing the phone case model into Cura LulzBot Edition

As I have done in the past few practical application chapters, I've imported my case model into *Cura LulzBot Edition* to prepare it for production for my 3D printers. Not every 3D printer is able to print flexible materials such as TPU, as was discussed in *Chapter 10*. For **FFF**-style printers, you typically need a printer that can print at higher temperatures, and ones equipped with extruder assemblies and nozzles that are suitable for abrasive materials. There are also flexible resins available for resin-style printers, but not all resin printers can use them. It's important to research what your printer can do before moving forward, and remember that you can always take advantage of 3D printing services, as discussed in *Chapter 14*, should you need a part made using something that you don't have readily available.

Additionally, there may be some preparation steps needed to print with these flexible materials, but before moving on to printing the case design, there are a few important settings to consider in our CAM program:

- **Material** – Flexible materials such as TPU come in many different forms and are typically rated on how flexible they are. Lower numbers, such as 75, would represent a more flexible material than a higher rating, such as 95. The more flexible the material, the slower you typically need to print it. If your printer can print with flexible materials, the print speed and printing temperatures should all be adjusted for you when selecting them in your slicer program. In general, flexible materials print at half the speed of rigid ones, and usually at 20–30 degrees hotter than PLA for nozzle temperatures on FFF-style printers.

- **Quality** – Flexible materials retain less detail than rigid materials and print better with thicker layer heights. Standard or high-speed printing profiles would typically work well, and better than a high-detail profile with a layer height of less than 2 mm.

- **Infill Density** – Flexible materials typically do better with a slightly higher infill density to balance temperatures and adhesion across layers. But as our phone case model is rather thin, there will be very little infill across the whole model. A density of 25% should do well to support this project with flexible materials.

- **Supports** – As previously mentioned, support material is very difficult to remove from flexible materials. More rigid TPUs, such as ones rated for 90 or 95, may do better with removable supports. But in general, we should avoid using supports with this type of material. The exception would be if you were working with a multi-extruder printer, as discussed in *Chapter 10*, which would allow for a different material to be used to print the supports.

After adjusting our print settings based on our selected material, we can slice and preview the model to check the first layer for any possible issues, as shown in *Figure 20.27*:

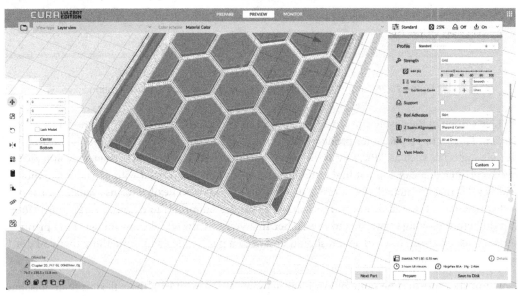

Figure 20.27: Inspecting the first layer of the prepared print

I also like to increase the number of lines printed for my **skirt**, as shown in *Figure 20.27*. This will trace the outline of my model a few more times, which allows me to check the calibration and settings to ensure optimal print performance when working with this specialty material.

Before we move ahead to printing the flexible material, your printer may require additional setup steps to find success. For example, flexible TPUs typically print best on a glass bed coated with a PVP-based glue stick when printing with FFF-style printers, as shown in *Figure 20.28*:

Figure 20.28: Preparing the bed of an FFF-style printer for TPU

When changing your build platform or coating it with glue, as shown in *Figure 20.28*, you may also need to adjust your offsets and calibration. Adjusting the number of lines for your skirt, as discussed earlier, makes this easier to do on the fly as you start your print.

Once your model and printer have been prepared properly, you can move on to manufacturing your phone case, as shown in *Figure 20.29*:

Figure 20.29: 3D printing the case model using flexible TPU filament on an FFF-style printer

Once the print has finished, you may find that there are some strings and over-extruded areas that need to be cleaned up, as shown in *Figure 20.30*:

Figure 20.30: Inspecting the printed case model

These can carefully be pulled or cut away using tweezers or scissors as needed. Once done, you can test the fit of your case with your mobile device! As with the previous chapters, remember that this is a prototype and there may be some improvements needed to find success. This is also a rather difficult challenge as there are many crucial measurements and tolerances that need to be accounted for. Don't be discouraged if you need to make a few adjustments and reprint before you have the perfect fit, as shown in *Figure 20.31*:

Figure 20.31: A successful 3D-printed phone case alongside failed iterations

But after some trial and error, you should be able to find success as you create an aesthetically pleasing case for your chosen mobile device, as shown in *Figure 20.31*. Again, while this chapter outlined creating a case for a phone, the skills and tools utilized are universally applicable when designing a 3D-printed prototype around the constraints of an existing device. From here, the possibilities of what we have covered are truly endless!

Summary

As this book is now coming to a close, it's important to not only consider the key topics and skills discussed in this chapter but also how they connect to our overall objective for this text. As demonstrated in this practical application, you now have the skills and abilities to design and create a real-world product, such as a phone case.

With your skills in taking measurements, planning out a project, utilizing CAD to design models, testing and simulating the functionality of these models, preparing them for manufacturing, and manufacturing them successfully using rapid prototyping technology, truly anything is achievable.

While it may not always be immediately clear, many of the strategies and tools implemented across this project and the ones completed previously are universal and can be utilized for other projects as well. Being able to implement dimensions, develop custom shapes, adjust shape parameters, and work through creating complex multi-part models is what makes for an effective CAD user. Even more so, having an understanding of how things are actually made and brought into the physical world, and then being able to design effective solutions for manufacturing is not always a universal skill, but an important one to have nonetheless.

And it's also important to remember that 3D printers come in many forms, but no one type of printer is perfect for every situation you may find yourself in. Printing simple boxes and things is rather easy to do, but changing your material, striving for tighter tolerances, making tiny things, or making larger things that print for more than a day is where everything starts to become a bit more challenging. Take time to learn, make mistakes, and don't be discouraged when something breaks!

As you begin to plan and implement practical applications of your own, remember to always consider the limitations of your resources in your design, and then work to design effective solutions around these limitations. This will allow you to take full advantage of what you might have available and create prototypes that are more effective and successful. Always consider aesthetics and ergonomics too, which will allow your projects to be more enjoyable to use, and remember that there is never one way to solve any problem!

But most importantly, make lots of mistakes, try to learn from them, and never stop tinkering.

Index

Symbols

2D orthographic views 38-40
3D Design 9
   Blocks mode 12
   Bricks mode 11
   Sim Lab mode 9, 11
3D model
   importing 116-118
3D objects
   importing 116
3D-printed phone case
   aesthetic and ergonomic features, adding 380
   corners, rounding 369-371
   dimensions, acquiring 366, 367
   making 375
   manufacturing, with specialty materials 384-389
   modeling 367, 368
   opening, creating 378-380
   phone model, copying 375, 376
   ports and buttons, adding 371-375
   ports and buttons, creating 376, 377
   protective lip, adding 380-383
   unique design features, adding 383

3DprinterOS 241
3D printing 146-148
   frequently used materials 160, 161
   material, selecting 157, 158
   working 149-151
3D printing services
   finding 241-243
3D printing techniques
   Fused Deposition Modeling (FDM) 154
   Fused Filament Fabrication (FFF) 152, 153
   Stereolithography 156
   vat photopolymerization 154-156
3D printing Tinkercad designs
   preparing 232-240
   preparing, with Autodesk Fusion 229
3D puzzle
   artwork, adding 331-335
   exporting 336-340
   joints, adding 327-331
   joint template, making 325-327
   manufacturing 336-340
   pieces, making 320-324
   preparing 336-340

Index

A

acrylonitrile butadiene styrene (ABS) 160
additive manufacturing 146
Align tool
    using 95-100
artwork, 3D puzzle
    adding 286, 331-336
    image features, adding 288, 289
    text features, adding 286, 287
AR Viewer 24
assembly system, catapult
    assembly beams, creating 350-353
    assembly holes, creating 347-350
    creating 347
Astroprint 241
Autodesk Fusion 229
    3D printing Tinkercad designs 229
    URL 216

B

Bambu Lab 241
base part, trophy
    base platform, creating 258-260
    connection point, creating 260, 261
    recipient, adding 262-264
Blocks mode, 3D Design 12
Braille shape generator 69
Bricks mode, 3D Design 11
brims 178
build plate
    optimizing 179-182

C

calipers 21
CAM program 385
CAM software 166
    selecting 218, 219
catapult
    assembly system, creating 347
    frame, designing 342
    manufacturing 359, 360
    projectile system, creating 353
    prototype testing 359-363
catapult frame
    Box shape, starting with 343, 344
    designing 342, 343
    shapes, adding to 344-346
    technical drawing, creating 342
chamfers
    creating 171, 173
    creating, with constructive solid
        geometry (CSG) 174, 175
Circuit 13, 14
clearance fit 189, 283
Codeblocks 14, 15, 137
    model, exporting from 16
computer-aided design (CAD) 3
Computer-Aided Manufacturing (CAM) 150
constructive solid geometry (CSG) 42-45
    using, to create fillets and chambers 174, 175
cruise mode 88
Cruise tool 92
    using 92-95
Cura
    adjusted settings, applying 227-229
    design, preparing 222
    print settings, adjusting 223-226
    setting up 219-221
    URL 216
Cura LulzBot Edition 218
curved words shape generator 70, 71
custom shapes
    creating 140, 141

D

designs
 exporting 216-218
 tinkering, to another 113-116
device
 selecting 23, 24
dimensions 21, 77
Draw Shape tool 130
Draw tool 129, 130
Duplicate and repeat tool 100, 182
 using 101-109

E

electronic circuits 13
endpoint 84
ergonomic features
 adding 309
 jar part, improving 309-312
 lid part, improving 312, 313
extruder 152
extrusions 62, 63
extrusion-type method 62
extrusion-type printers
 working with 208-210

F

FFF-style printers 385
filament 152
fillets
 creating 171, 173
 creating, with constructive solid geometry (CSG) 174, 175
first layer
 designing 175, 176
 rafts and brims, adding 178, 179
 surface area, utilizing 176, 177

fit
 testing, with socket and peg shapes 192-194
 types, considering 188, 189
formative manufacturing 147
Fused Deposition Modeling (FDM) 154, 186
Fused Filament
  Fabrication (FFF) 152, 153, 185, 186
 benefits 153

G

Gcode file 151, 228, 269, 317
Graphics Library Transmission Format 217
grid
 working with 78-82
grooves
 cutting 279, 280-283
grouping shapes technique 45
groups
 models, creating 48, 49
 shapes, adding 49-51
 working in 45-48

I

imported 3D model
 manipulating 119, 120
individual characters
 using 67
isometric drawing 20, 38

J

jar
 exporting 314-318
 manufacturing 314-318
 modeling 298-302

joints, 3D puzzle
  adding 327-331
joint template, 3D puzzle
  making 325-327

K

keyboard shortcuts 26-29
Kiri:Moto
  URL 216

L

layer height 206
lid
  creating 283-285
  modeling 302, 303

M

manufacturing techniques
  comparing 146
material properties 9
measurements
  obtaining 21-23
microcontrollers 13
midpoint 84
Mirror tool 66, 96
models
  exporting 266-270
  manufacturing 270-273
  preparing for production 266-270, 290-294
mouse
  using 25, 26
multi-part storage box 276
  aligning 279
  dimensioning 277
  starting with 277

multiple sides
  viewing 32-35

O

Octoprint 241
orbiting 25, 33, 35
orthographic drawing 20, 38
Overhang 63, 164, 166, 352
  avoiding 164-169

P

pan 36
parametric approach 128
parametric modeling 43
parts
  connecting 212, 213
peg shape
  creating, to fit into socket shape 191, 192
perspective views
  orbiting through 35-37
pockets 64
polyethylene terephthalate
  glycol (PETG) 160
Polylactic Acid (PLA) 158
  printing with 158-160
press fit 189
print beds 152
print farms 148
projectile system, catapult
  creating 353
  frame, completing 354-356
  making 358-363
  projectile arm, making 356-358
Protolabs 242
prototyping 147

R

rafts 178
rapid-prototyping 147
resin printing 154
resin-type 3D printers
  working with 210
rough sketching 18
rulers 21
ruler tool 82
  using 83-85

S

Scribble shapes 128
  basics 128
  Draw Shape tool 130
  Draw tool 129, 130
  editing tools 130, 131
Scribble tool
  using 131-136
script shape generator 68
segments
  adjusting 170
  creating 169
Shape Generators 136
  writing 137-139
shape parameters
  manipulating 51-53
shapes
  aligning 95-100
  duplicating 100-109
  hiding 53, 54
  locking 53, 54, 55
  patterning 100-109
shape transformations 101
Shapeways 242
Sim Lab mode, 3D Design 9, 11
SimplyPrint 241
sketch
  working with 18-21
Slic3r
  URL 216
slicers 151, 202, 216
snap grid 182
socket shape
  creating 189, 190
Stereolithography (SLA) 156, 186, 217
strategies, for creating models for 3D Printing
  chamfers, creating 169
  fillets, creating 169
  first layer, designing 175, 176
  overhangs, avoiding 164-166
  segments, creating 169
subtractive manufacturing 147
support material 166
SVG images 128

T

technical drawing 19
text
  grouping 60, 61
  typing 60, 61
text orientation 64-66
text ring shape generator 70
text shape generators
  Braille shape generator 69
  curved words shape generator 70, 71
  script shape generator 68
  text ring shape generator 70
  using 68
text shapes
  example shape 71-76
thermoplastic polyurethane (TPU) 160

thin lines and walls
    identifying 206-208
    performance, adjusting 210, 211
threads
    modeling 304-308
thumbnail sketches 18
Tinkercad 3
    3D Design 8, 9
    Circuit 13, 14
    Codeblocks 14-16
    creating 5, 6
    fits, modeling 189
    URL 26
    users 6-8
Tinkercad designs
    sending, to Fusion 230-232
    tolerances, adding to 188
tolerances 184, 185, 283
    adding, to Tinkercad designs 188
    additional factors, for determining accuracy 187, 188
    applying, in real-world setting 195-199
    calculating 186
    determining, based on material option 187
    determining, based on printer type 186
    type of fit, considering 188, 189
top part, trophy
    cup, modeling 248-250
    cup part, finishing 254-257
    designing 248
    post, modeling 250-253
transition fit 189
tree supports 270
trophy
    base part, designing 257
    parts, reorienting 264-266
    top part, designing 248

V

vat photopolymerization 154, 156
vector shapes
    custom vector image files, creating 121-123
    custom vector image files, importing 121
    importing 120
    web, browsing for vector image files 124-126
view cube 35, 36

W

Wavefront OBJect 217
workplane 78-81, 92
    using 85-88
    watching 202-206
workplanes

X

Xometry 242

Z

zooming 36

packtpub.com

Subscribe to our online digital library for full access to over 7,000 books and videos, as well as industry leading tools to help you plan your personal development and advance your career. For more information, please visit our website.

Why subscribe?

- Spend less time learning and more time coding with practical eBooks and Videos from over 4,000 industry professionals
- Improve your learning with Skill Plans built especially for you
- Get a free eBook or video every month
- Fully searchable for easy access to vital information
- Copy and paste, print, and bookmark content

Did you know that Packt offers eBook versions of every book published, with PDF and ePub files available? You can upgrade to the eBook version at packtpub.com and as a print book customer, you are entitled to a discount on the eBook copy. Get in touch with us at customercare@packtpub.com for more details.

At www.packtpub.com, you can also read a collection of free technical articles, sign up for a range of free newsletters, and receive exclusive discounts and offers on Packt books and eBooks.

Other Books You May Enjoy

If you enjoyed this book, you may be interested in these other books by Packt:

Practical Autodesk AutoCAD 2023 and AutoCAD LT 2023

Jaiprakash Pandey, Yasser Shoukry

ISBN: 978-1-80181-646-5

- Understand CAD fundamentals like functions, navigation, and components
- Create complex 3D objects using primitive shapes and editing tools
- Work with reusable objects like blocks and collaborate using xRef
- Explore advanced features like external references and dynamic blocks
- Discover surface and mesh modeling tools such as Fillet, Trim, and Extend
- Use the paper space layout to create plots for 2D and 3D models
- Convert your 2D drawings into 3D models

A BIM Professional's Guide to Learning Archicad

Stefan Boeykens, Ruben Van de Walle

ISBN: 978-1-80324-657-4

- Create an architectural model from scratch using Archicad as BIM software
- Leverage a wide variety of tools and views to fully develop a project
- Achieve efficient project organization and modeling for professional results with increased productivity
- Fully document a project, including various 2D and 3D documents and construction details
- Professionalize your BIM workflow with advanced insight and the use of expert tips and tricks
- Unlock the geometric and non-geometric information in your models by adding properties and creating schedules to prepare for a bill of quantities

Packt is searching for authors like you

If you're interested in becoming an author for Packt, please visit `authors.packtpub.com` and apply today. We have worked with thousands of developers and tech professionals, just like you, to help them share their insight with the global tech community. You can make a general application, apply for a specific hot topic that we are recruiting an author for, or submit your own idea.

Share Your Thoughts

Now you've finished *Taking Tinkercad to the Next Level* we'd love to hear your thoughts! Scan the QR code below to go straight to the Amazon review page for this book and share your feedback or leave a review on the site that you purchased it from.

`https://packt.link/r/<1835468004>`

Your review is important to us and the tech community and will help us make sure we're delivering excellent quality content.

Download a free PDF copy of this book

Thanks for purchasing this book!

Do you like to read on the go but are unable to carry your print books everywhere?

Is your eBook purchase not compatible with the device of your choice?

Don't worry, now with every Packt book you get a DRM-free PDF version of that book at no cost.

Read anywhere, any place, on any device. Search, copy, and paste code from your favorite technical books directly into your application.

The perks don't stop there, you can get exclusive access to discounts, newsletters, and great free content in your inbox daily

Follow these simple steps to get the benefits:

1. Scan the QR code or visit the link below

https://packt.link/free-ebook/9781835468005

2. Submit your proof of purchase
3. That's it! We'll send your free PDF and other benefits to your email directly

Made in the USA
Monee, IL
21 April 2025

16116555R00230